Design and Control of Grid-Connected Photovoltaic System

The current model for electricity generation and distribution is dominated by centralized power plants which are typically associated with combustion (coal, oil, and natural gas) or nuclear generation units. These power models require distribution from the center to outlying consumers and have many disadvantages concerning electric utilities, transmission and distribution, and greenhouse gas emissions. This resulted in the modeling and development of cleaner renewable power generation with alternative sources such as photovoltaic (PV), wind, and other sources. Further, due to matured PV technology, constant drop-in installation cost, greenhouse emissions reductions, energy efficiency, reduced transmission and distribution investments, minimization of electric losses, and network support, the development of PV systems is proliferating. In view of this development, this book provides an idea for setting up the PV plant from the initial study of the site to plan sizing. Once the first planning is covered, then the book focuses on the modeling aspects of power electronics converter and control elements associated with it keeping the operating standards specified for the development of distributed generation systems in check.

This book will be useful for industrial professionals and researchers who are working toward modeling of PV plants, and their control in grid connected operation. All the necessary information related to these fields is available in the book.

Power Electronics and Applications Series

Series Editor:
Muhammad H. Rashid, University of West Florida, Pensacola, USA

Complex Behavior of Switching Power Converters
Chi Kong Tse

DSP-Based Electromechanical Motion Control
Hamid A. Toliyat and Steven Campbell

Advanced DC/DC Converters
Fang Lin Luo and Hong Ye

Renewable Energy Systems: Design and Analysis with Induction Generators
M. Godoy Simoes and Felix A. Farret

Uninterruptible Power Supplies and Active Filters
Ali Emadi, Abdolhosein Nasiri, and Stoyan B. Bekiarov

Modern Electric, Hybrid Electric, and Fuel Cell Vehicles: Fundamentals, Theory, and Design
Mehrdad Eshani, Yimin Gao, Sebastien E. Gay, and Ali Emadi

Electric Energy: An Introduction
Mohamed EI-Sharkawi

Electrical Machine Analysis Using Finite Elements
Nicola Bianchi

Design and Control of Grid-Connected Photovoltaic System
Ahteshamul Haque, Mohammed Ali Khan, and Varaha Satya Bharath Kurukuru

For more information about this series, please visit: https://www.routledge.com/ Power-Electronics-and-Applications-Series/book-series/CRCPOWELEAPP
NICOLA BIANCHI

Design and Control of Grid-Connected Photovoltaic System

Ahteshamul Haque, Mohammed Ali Khan, and Varaha Satya Bharath Kurukuru

CRC Press
Taylor & Francis Group
Boca Raton London New York

CRC Press is an imprint of the
Taylor & Francis Group, an **informa** business

First edition published 2023
by CRC Press
6000 Broken Sound Parkway NW, Suite 300, Boca Raton, FL 33487-2742

and by CRC Press
4 Park Square, Milton Park, Abingdon, Oxon, OX14 4RN

CRC Press is an imprint of Taylor & Francis Group, LLC

ISBN: 9781032189741 (hbk)
ISBN: 9781032189772 (pbk)
ISBN: 9781003257189 (ebk)

DOI: 10.1201/9781003257189

Typeset in Times
by codeMantra

Contents

Authors

Mohammed Ali Khan (S'17-M'22) is currently working as a post-doctoral researcher at the Centre for Industrial Electronics, University of Southern Denmark, Sønderborg, Denmark. He received his B.Tech. degree in Electrical and Electronics Engineering from Karunya University, Coimbatore, India, in 2013, and M.Tech. degree in Power System from Amity University, Noida, India, in 2016. He did Ph.D. in Power Management of Grid Connected Distribution Generation at Advanced Power Electronics Research Laboratory, Department of Electrical Engineering, Jamia Millia Islamia (A Central University), New Delhi, India, in 2021. He had also worked as post-doctoral researcher in the Department of Electrical Power Engineering, Faculty of Electrical Engineering and Communication, Brno University of Technology, Brno, Czech Republic from December 2021 to January 2023. He was a visiting researcher at the Center of Reliable Power Electronics, Aalborg University, Aalborg, Denmark, from October to December 2020. He had also worked as Guest Faculty in the Department of Electrical Engineering, Jamia Millia Islamia (A Central University), New Delhi, India from 2017 to 2020. He has numerous publications in peer-reviewed journals and presented his research articles in several international conferences. His area of research is artificial intelligence, power electronics, and their application in renewable energy systems, power quality improvements, and reliability.

Ahteshamul Haque (M'13-SM'14) received a B.Tech. degree in Electrical Engineering from Aligarh Muslim University, Aligarh, India, in 1999, a Master's degree in Electrical Engineering from IIT Delhi, New Delhi, India, in 2000, and a Ph.D. degree in Electrical Engineering from the Department of Electrical Engineering, Jamia Millia Islamia University, New Delhi, India, in 2015. Prior to academics, he worked in the research and development unit of world-reputed multinational industries, and his work is patented in the USA and Europe. He is currently an associate professor at the Department of Electrical Engineering, Jamia Millia Islamia University. He has established the Advance Power Electronics Research Laboratory, Department of Electrical Engineering, Jamia Millia Islamia. He is working as a principal investigator of the MHRD-SPARC project and other research and development projects.

He is the recipient of the IEEE PES Outstanding Engineer Award for the year 2019. He has authored or co-authored around 100 publications in international journals and conference proceedings. He is a senior member of the IEEE. His current research interests include power converter topologies, control of power converters, renewable energy, and energy efficiency, reliability analysis, electric vehicle operations.

Varaha Satya Bharath Kurukuru (S'18-M'22) received his B.Tech. degree in Electrical and Electronics Engineering from Avanthi's Research and Technological Academy, Vizianagaram, India, in 2014, and M.Tech. degree in Power Systems from Amity University, Noida, India, in 2016. He completed his Ph.D. degree in Intelligent Monitoring of Solar Photovoltaic System from the Advanced Power Electronics Research Laboratory, Department of Electrical Engineering, Jamia Millia Islamia (A Central University), New Delhi, India, in 2021. He is currently working as a scientist (packaging and multiphysics) at Power Electronics Research Division, Silicon Austria Labs GmbH, Villach, Austria.

He was a visiting researcher at the Center of Reliable Power Electronics, Aalborg University, Aalborg, Denmark, from August to October 2019. His area of research is thermal characterization of packaged power devices, condition monitoring, and reliability of power electronic converters in renewable energy systems and electric vehicles.

1 Site Study for Grid-Connected PV Systems

1.1 INTRODUCTION

Over the past decade, the rise of renewable energy sources in the power sector has limited carbon emissions and avoided a 10% increment in carbon emissions every year. In 2020, the share of renewables in global electricity generation has increased from 27% to 29%27% in 2019. This has resulted in a decline in carbon emissions by 3.3% in 2020 [1]. The need to decarbonize the power sector has led to a rise in the installation of renewable distribution generation (DG) systems globally. As per the report of the International Renewable Energy Agency (IRENA) [2], more than 260 GW of renewable energy capacity was installed in 2020. More than 80% of all electricity capacity installed was renewable and particularly solar and wind as they accounted for 91% of newly installed renewable energy sources. The statistics of variation in global energy generation are shown in Figure 1.1.

The growth rate of solar energy from 2019 to 2020 has been 20.5%, whereas the wind growth rate has been 11.9%, and it is 5.3% for the other renewable sources [3]. The shift toward renewable energy sources has been intensifying in the recent years to achieve carbon neutrality. As per the report of the IRENA [4], about 66% of the total primary energy supply will be from renewable sources by 2050 which will result in the decline of the carbon intensity of the power sector by 85%.

For installation of solar DG systems, multiple constraints such as the location of the panel, the capacity to be installed, irradiance received during the day at the site, etc. need to be considered [5]. The site selection is very important for solar DG systems as the efficiency of the system will be impacted based on the site location. The capacity of solar DGs can vary for a large-scale plant to a rooftop solar installation. Site is selected based on the irradiance received throughout the day as the irradiance is directly proportional to the power generated for the panel. For a rooftop solar installation, the condition of the roof, its slope, and the weight it can withstand needs to be considered. In case of a rooftop solar installation, the condition of the roof, its slope, and the weight it can withstand needs to be considered. The load demand that has to be satisfied by the PV system needs to be analyzed for identifying the panel capacity which is to be installed. Different countries have different building, electrical, and fire codes which also need to be considered while installing solar panels.

1.2 DESIGN AND SIZING PRINCIPLES

Appropriate system design and component sizing are fundamental requirements for reliable operation, better performance, safety, and longevity of solar photovoltaic

GLOBAL ENERGY GENERATION MIX

■ Coal   ■ Gas   ■ Nuclear   ■ Renewable

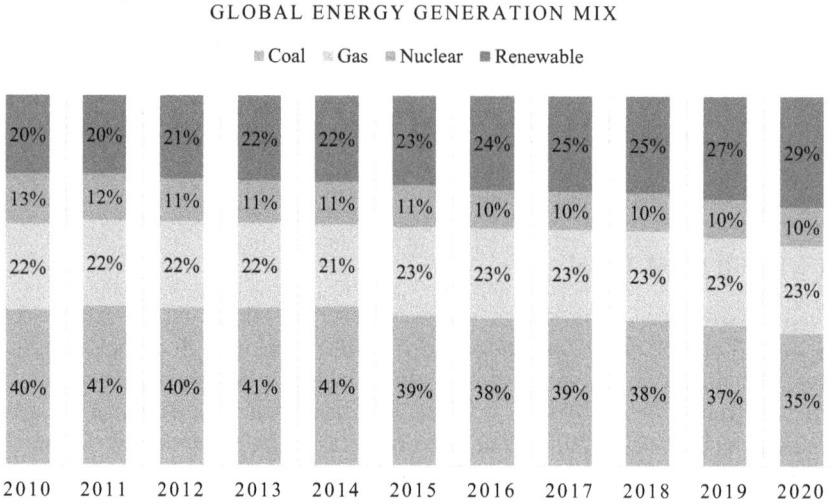

2010	2011	2012	2013	2014	2015	2016	2017	2018	2019	2020
20%	20%	21%	22%	22%	23%	24%	25%	25%	27%	29%
13%	12%	11%	11%	11%	11%	10%	10%	10%	10%	10%
22%	22%	22%	22%	21%	23%	23%	23%	23%	23%	23%
40%	41%	40%	41%	41%	39%	38%	39%	38%	37%	35%

FIGURE 1.1 Global energy generation based on the source of energy generation.

(PV) systems. Hence, to achieve an efficient operation with the sizing of the PV system components, it is necessary to identify the factors that can affect their performance [6,7].

1.2.1 FACTORS AFFECTING PV MODULE PERFORMANCE

The performance of a PV module is directly related to the amount of sunlight it receives. If the PV module is even partially shaded, the performance is very low. Apart from the shading phenomenon, there are many other factors that affect the performance of the PV system. These factors must be understood in order for clients to have realistic expectations of overall system performance and economic benefits in weather conditions that change over time [8,9].

Location: The starting point when planning a PV system is the selection of location. The amount of solar access a PV module gets is very important to the financial viability of a PV system. Besides, latitude is considered the main factor for identifying a location [10].

Solar Irradiance: This is a measure of how much sunlight intensity or power, expressed in wattage per square meter, hits a flat surface or can reach your location. As the weather conditions are relatively similar over the years, historical standardized weather data can be used to predict the average monthly and yearly energy production of the system. There is a map of solar resources showing the amount of energy reaching the surface of the module. The data are displayed on a standardized map, showing how many hours of standard daylight hours can be determined in a month or year. This is referred to by the term "solar radiation."

Solar Insolation: The amount of solar insolation is a measurement of the amount of solar irradiance that reaches the PV surface at a specific point in time. The solar energy available at a particular location is expressed in kWh/m^2/day. This is commonly known as the peak sun hour (PSH). For example, if the amount of solar radiation at a location is 5 kWh/m^2/day, the PSH at that location is hours. Installing a 1 kW solar panel in this location will generate 1 kW $\times$ 5h = 5 kWh of energy per day without considering the loss. The stronger the sunlight, the higher the output of the module. If the amount of solar radiation is low, the power output will also be low. The change in voltage due to the fluctuation of the amount of solar radiation is not noticeable.

The map in Figure 1.2 shows the amount of solar energy available (in hours) to generate electricity daily on the optimal slope for the worst month of the year (based on accumulated total solar radiation data). This is very useful because it can be adopted to calculate the energy production of the solar system.

Statistical estimates of average daily solar radiation at a particular location are often used in the PV design process and are measured in kilowatt-hours $\left(\text{kWh}/m^2/\text{day}\right)$ per square meter per day.

Electricity Generation vs the Sun Hours Available per Day: Several factors affect the amount of solar energy that a module is exposed to, and they are listed as follows:

- When to use the system – summer, winter, or all year round.
- Typical regional weather conditions.
- Fixed mount pair vs trackers.
- Position and angle of PV generator.

Air Mass: Air mass refers to the "thickness" and transparency of the air that sunlight reaches the module (the solar altitude affects this value). The standard is 1.5.

Sun Angle and PV Orientation: The direction in which the solar panel is facing is referred to as its orientation. The orientation of the solar array is very important because it affects the amount of sunlight that hits the solar array and thus the amount

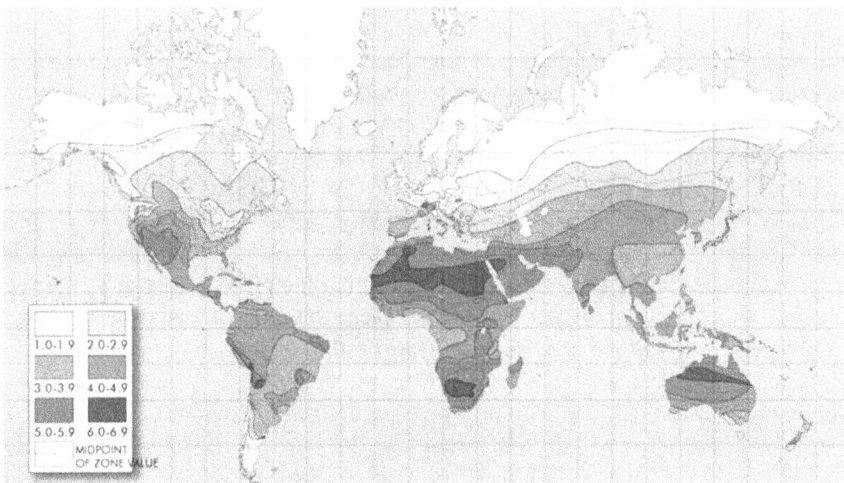

FIGURE 1.2 Map of solar energy available to generate electricity [11].

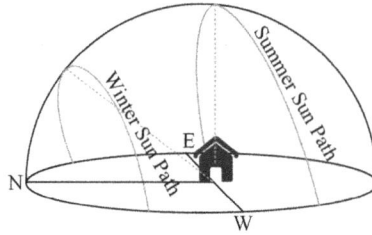

FIGURE 1.3   The changing path of the sun [12].

of energy that the solar array produces. Orientation generally includes the direction the solar module is facing (i.e., just south) and the tilt angle, which is the angle between the base of the solar module and the horizon. The amount of sunlight that hits the array also depends on the time of day, depending on the movement of the sun across the sky. The sun's path during summer and winter is shown in Figure 1.3. Solar modules should be installed so that as much radiation as possible is collected. Ideally, the PV system at the north of the equator works best when it is directed south and tilted at an angle of 15° above the latitude of the location. Installing a PV generator in a building where the module is unlikely to face south can be east-facing or west-facing, but not north-facing, as the efficiency is very limited. The highest efficiency or peak performance of a PV module occurs when its surface is perpendicular to solar radiation. As the rays deviate from the vertical direction, more energy is reflected by the module rather than being absorbed.

Most PV systems are permanently installed and cannot follow the sun all day long. Performance can be improved by installing PV modules on the tracker to follow the sun from east to west (single-axis tracker) during the day and from north to south (biaxial tracker) during seasonal changes. This can be expensive and is not common in most PV applications.

The PV needs to be tilted toward the average altitude of the sun, which is the latitude of the array location, in order to capture most of the solar energy for the year. For example, at 25° latitude, a system that is used all year round can be tilted between 15° and 35° to maximize the amount of electricity for a year.

Shading: Shading is one of the most important parameters of energy loss in PV systems. Even partial shading of one cell in a 36-cell module can significantly reduce output. Potential causes of shade can be self-shading from multiple rows of trees and bushes, adjacent buildings, and the module itself. To avoid the shade of noon in winter, calculations should be performed to determine the minimum spacing between columns of the PV array. As a rule of thumb, place the array at least twice the height of the object. This will prevent the object from casting shadows on either side of noon for 4 hours. As a general rule, the smaller the tilt angle, the less the shade you have and the more effectively you can use the area. However, after that, the amount of solar power generation will decline throughout the year. For this reason, a tilt angle of 15° is usually selected. Thin-film PV modules are more resistant to partial shading than crystalline silicon PV modules.

1.2.2 ELECTRICAL CHARACTERISTICS

The solar cell converts sunlight to "DC" or direct current electricity, which is the same type of electricity that is produced by every-day batteries where electrons flow in one direction. Most PV solar cells generate an open-circuit voltage of approximately $0.5 - 0.6\,\text{V}$ when no external circuit is connected. This output voltage (V_{OUT}) is highly dependent on the PV cell load current requirement. For example, on a very cloudy day, the lower power requirements allow the cell to provide the maximum output voltage V_{OUT}, but the output current is reduced. However, as the current demand of the load increases, brighter light (solar radiation) is needed at the junction to maintain the maximum output voltage V_{OUT}. However, no matter how strong or bright the solar radiation is, there is a physical limit to the maximum current that a single solar cell can supply. This is called the maximum available current and is expressed as I_{MAX}. The I_{MAX} value of an individual solar cell depends on the size or surface area of the cell, the amount of direct sunlight that hits the cell, the efficiency of converting that solar energy into electricity, and of course the type of semiconductor material. Cells will be made from silicon, gallium arsenide, cadmium sulfide, cadmium telluride, etc. Most commercial PV solar cells have a nominal PV value that is indicated by P_{MAX}, the largest available PV that the cell can supply. This corresponds to the product of the cell voltage V multiplied by the maximum cell current I and is given as:

$$P_{MAX} = V_{OUT} \times I_{MAX} \qquad (1.1)$$

where P is the power in watts, V is the voltage in volts, and I is the current in amperes. Various manufacturers refer to the total solar cell output of a solar cell as the "maximum output," "peak power," "nominal power," "maximum power point," etc., all of which have the same meaning.

Photovoltaic IV Characteristic Curves: PV solar cell manufacturers emulate a current–voltage (IV) curve that shows the current and voltage at which a PV cell produces maximum output, and the cell is shadowless under standard conditions of sunlight and temperature. The typical current–voltage curve of a PV module is shown in Figure 1.4.

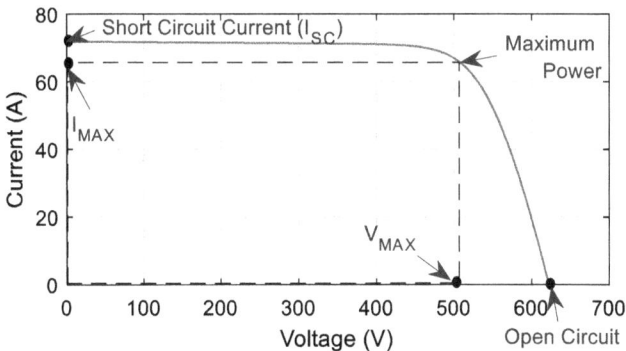

FIGURE 1.4 Typical current–voltage curve of a photovoltaic module.

Voltage (V) is plotted on the horizontal axis and current (I) is plotted on the vertical axis. The available power (W) of PV at each point on the curve is the product of current and voltage at that point.

Short-Circuit Current (I_{SC}): Photovoltaic modules generate maximum current when the circuit is essentially free of resistance. This is a short circuit between the positive and negative poles. This maximum current is known as the short-circuit current (I_{sc}). This value is higher than I_{max}, which represents the normal operating current of the circuit. In this state, the resistance is zero and the circuit voltage is zero.

Open-Circuit Voltage (V_{OC}): The open-circuit voltage (V_{oc}) means that no current is flowing because the PV cell is not connected to an external load (open-circuit state). This value depends on the number of PV modules connected in series. In this state, the resistance is infinitely high and no current flows.

Maximum Power (P_{MAX} or MPP): This is the point at which the power of the array (battery and inverter) connected to the load reaches its maximum, where $P_{max} = I_{max} \times V_{max}$. The maximum power point of a PV system is measured in watts (W) or peak watts (Wp). The I_{max} and V_{max} values are on the "knee" of the IV curve.

Fill Factor (FF): The FF is the ratio of the product of the maximum power output (P_{max}) to the open-circuit voltage and the short-circuit current $(V_{oc} \times I_{sc})$. The relationship is as follows:

$$FF = \frac{P_{max}}{V_{oc} \times I_{sc}} = \frac{I_{max} \times V_{max}}{V_{oc} \times I_{sc}} \tag{1.2}$$

Furthermore, the FF also provides an idea of the array quality. The closer the curve factor is to 1 (unit), the more the power the array can provide. Typical values are between 0.7 and 0.8.

PV Panel Energy Output: Previously, it was learned that the output of a solar cell is expressed in watts and is equal to the product of voltage and current $(V \times I)$. The optimum operating voltage for a PV cell under load is approximately 0.46 V at normal operating temperatures, producing approximately 33 amps of current when the sun is fully exposed. Next, the output of a typical solar cell can be calculated as follows:

$$P = V \times I = 0.46 \times 3 = 1.38 \text{ watts} \tag{1.3}$$

This may be fine for powering calculators, small solar chargers, or garden lamps, but these 1.38 watts are not enough to do a useful job. However, if the PV cells are connected in series (daisy chain), voltage is added, and if the PV cells are connected in parallel (adjacent to each other), the current is added. The right combination of series and parallel PV modules provides the required voltage, current, and power.

1.2.3 PV MODULE OUTPUT

At a given load, the performance of the PV module depends on two main factors: irradiance or light intensity and temperature.

Solar Intensity: The amount of sunlight that hits the front of the solar cell affects its performance. The more sunlight that passes through the cell, the more electricity it produces. The tension does not change. Figure 1.5a shows that under various test conditions, the output of the PV module changes proportionally when the sunlight is 1000 W/m^2 vs $600 W/m^2$. Dirt and dust can collect on the surface of the solar panel, blocking some of the sunlight and degrading performance. Although rigorous maintenance removes dust and dirt on a regular basis, it is more realistic to estimate system performance to account for the reduced accumulation of dust in the dry season. A typical annual dust reduction factor is 93% or 0.93. A "100-watt module" that operates on accumulated dust operates on average about 93% (85 watts × 0.93 = 79 watts).

Temperature: The performance of PV cells deteriorates as the cell temperature rises. The operating voltage decreases as the cell temperature rises. In full sun, the output voltage drops by about 5% for every $25°C$ increase in cell temperature. The

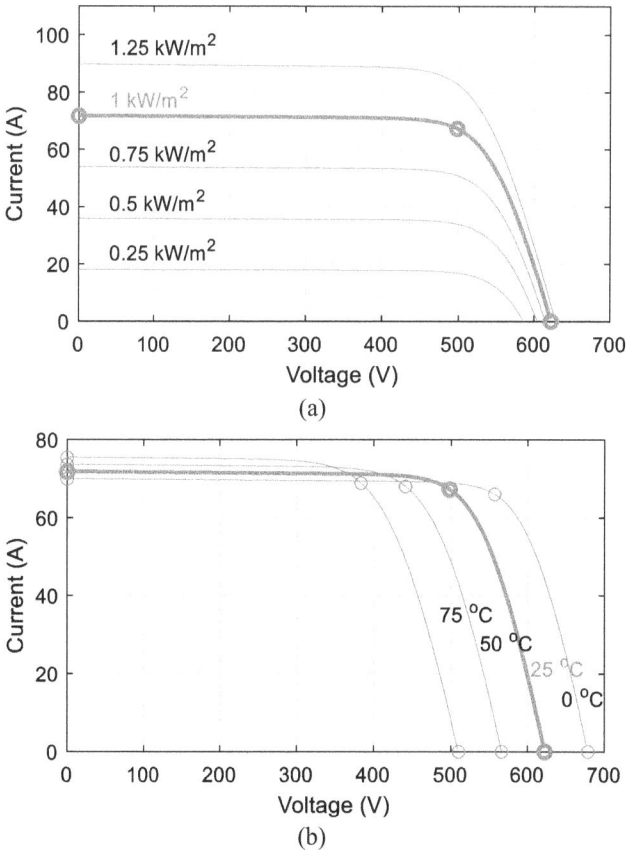

FIGURE 1.5 (a) IV curve at different light intensities. (b) IV curves at different temperatures.

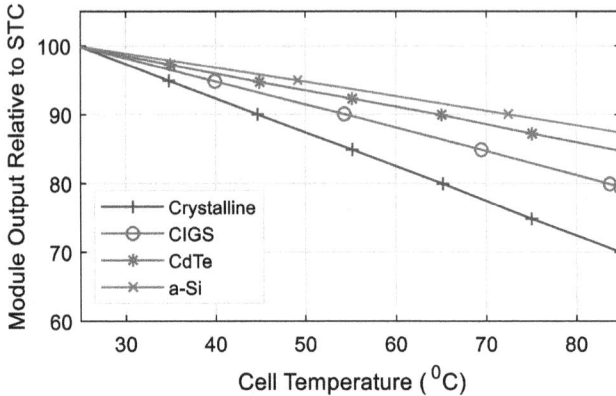

FIGURE 1.6   Effects of a negative temperature coefficient of power on PV module performance.

corresponding IV curves at different temperatures are shown in Figure 1.5b. To compensate for the power loss due to high temperatures, PV modules with more solar cells are recommended in very hot climates rather than colder ones.

Most thin-film technologies have a lower negative temperature coefficient than crystal technology. In other words, as the temperature rises, the loss of nominal capacity tends to decrease as shown in Figure 1.6. As the temperature of the solar cell rises, the open-circuit voltage V_{oc} decreases, but the short-circuit current I_{sc} increases slightly.

1.2.4   PV MODULE EFFICIENCY AND DERATING FACTORS

The efficiency of a PV module is the ratio of the electrical output power P_{out} compared to the PV input pin that hits the module. Solar cells can operate to maximum power to achieve maximum efficiency, so P_{out} can be considered P_{MAX}. The efficiency of a typical solar system is usually as low as about $10\% - 12\%$.

Example 1.1

On a clear sunny day, the 1 kWp PV system received 6 PSH. Estimate the expected output of the system.

Solution: Peak Power Output $\times$ Peak Sun Hours = Expected Output

$$1kW \times 6PSH = 6kWh$$

The above example indicates the maximum theoretical energy yield that is never produced in a real PV system. Due to the inefficiency and loss of the PV system, the actual output will be much lower than calculated. A summary of estimated losses can be found in Table 1.1.

Table 1.1
Summary of Estimated Losses

Cause of Loss	Estimated Loss (%)[a]	Derating Factor
Temperature	10%	0.90
Dirt	3%	0.97
Manufacturers tolerance	3%	0.97
Shading	2%	0.95
Orientation/tilt angle/azimuth	1%	0.99
Losses due to voltage drop in cables from the PV array to the battery	2%	0.98
Losses in distribution cables from the PV battery to loads	2%	0.98
Losses in a charge controller	2%	0.98
Battery losses	10%	0.9
Inverter	10%	0.9
Loss due to the irradiance level	10%	0.97
Total DF (multiplying all DFs)		0.60

[a] Typical loss of a PV system. The actual loss depends on the condition of the site.

Example 1.2

On a clear sunny day, the 1 kWp PV system received 6 PSH. Estimate the energy yield from the system.

Solution:

Energy Yield $=$ Peak Sun Hour $\times$ Module Rated Power $\times$ Total Derating Factor

$= 1 \text{ kWp} \times 6 \times 60\%$

$= 3.6 \text{ kWh}$

Performance Degradation Over Life Cycle: The performance of PV modules deteriorates over time as shown in Figure 1.7. The rate of deterioration usually increases during the first year of first exposure to light and then stabilizes. Factors that influence the degree of deterioration include the quality of the materials used in manufacturing, the manufacturing process, the quality of assembling and packaging the cells in the module, and on-site maintenance work. In general, the deterioration of high-quality modules is about 20% during the module life of 25 years, from 0.7% to 1% per year.

Example 1.3

On a clear sunny day, the 1 kWp PV system received 6 peak hours of sunshine. The total loss (derating factor) of the system is estimated to be 0.70 (70%). Estimate the expected output of the system.

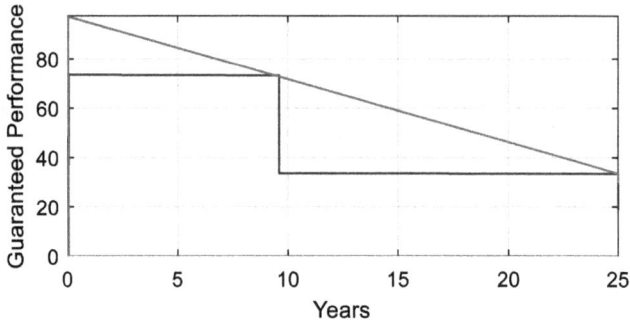

FIGURE 1.7 PV module degradation.

Solution:

Expected Output = Peak Sun Hours × Peak Power Output × Total derating factor

$= 1\text{kWp} \times 6 \text{ hours / day} \times 0.70$

$= 4.2$ kWh per day $\left(1\text{st year}\right)$

Now consider the deterioration of the module according to the indicator profile above (only as an example, the actual deterioration of the module is based on the quality of the module and the climatic conditions).

Energy generation:

$= 3.83$ kWh per day $\left(\text{on 10th year}\right)$

$= 3.39$ kWh per day $\left(\text{on 25th year}\right)$

1.2.5 SIZING PRINCIPLES

The sizing principles of a grid-connected, self-sufficient PV system are based on various design and functional requirements.

- Grid-connected systems without energy storage.
  - Providing additional power for system load.
  - There is no loss of load in the event of a PV system failure.
- Standalone system with energy storage.
  - Designed to meet specific electrical load requirements.
  - PV system failure leads to load loss.

PV Sizing: The equation that may be used to size a PV system is given by:

$$W_{\text{PV}} = \frac{E}{\text{PSH} \times \eta_{\text{sys}}} \tag{1.4}$$

where W_{PV} is the peak wattage of the array, E is the daily energy requirement, PSH is the average daily number of Peak Sun Hours in the design month for the inclination and orientation of the PV array, and η_{sys} is the total system efficiency.

The month in which the system is designed is the month with the lowest average daily solar radiation during system uptime. Peak hours are related to the tilt and orientation of the PV generator. If the only information available is about solar radiation in the horizontal plane, then tilt and direction correction factors should be applied.

System Sizing for Grid-Connected Systems: Sizing for grid connection systems without energy storage typically includes the following:

- Determines the maximum power output of the array.
- Based on available space, efficiency of PV modules used, array placement, and budget.
- Selection of one or more inverters with a total nominal power of 80%–90% of the maximum nominal power of the array with standard test conditions (STC).
- The sizing of the inverter string determines the specific number of modules connected in series that are allowed in each source circuit to meet the voltage requirements.
- The nominal power of the inverter limits the total number of parallel-source circuits.
- Estimates of system energy production based on local solar resources and meteorological data.
- The sizing of the interactive PV system is based on the requirements of the inverter.

The basic schematic for sizing of PV array in a grid-connected system is shown in Figure 1.8. The following steps will help you determine the array size of your grid-connected PV system.

Step 1: Find the monthly average electricity usage from the energy bill: This is the total kWh you pay in a month. Due to the seasonal use of air conditioners, heating, etc., it is advisable to check the invoice for several months of the year. Use all available data to determine your average monthly power consumption.

Step 2: Find the daily average electricity usage: Divide the monthly average kWh in step 1 by 30 days.

Step 3: Find the daily average PSHs for your location.

Step 4: Estimate the solar system size (AC) to generate 100% of your electricity consumption: divide the average daily energy expenditure (*step 2*) by the average peak time for the location. For example, if the average energy consumption is 34 kWh/day with a peak hour of sunshine 4.5 hours, the PV system size (AC) will be 34 kWh/4.5 h = 7.55 kW. Multiply it by 1000 to get watts.

Inverter size is
determined by
the PV array
maximum power

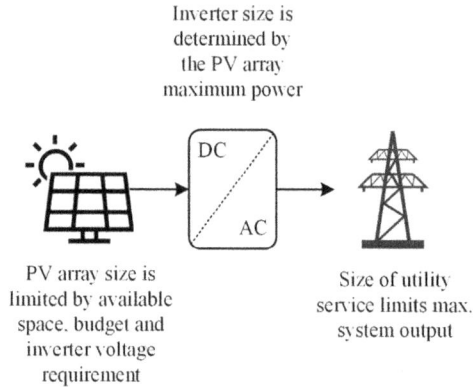

PV array size is
limited by available
space. budget and
inverter voltage
requirement

Size of utility
service limits max.
system output

FIGURE 1.8   Generalized schematic for sizing of the PV array in a grid-connected system.

Note: The size of the solar system calculated in step 4 is expressed in AC, which is the output of the solar system. Solar modules form the entrance to the solar system. Therefore, system inefficiencies must be taken into account when estimating the number of solar panels required. Traditional grid-connected photovoltaic systems lose about 14% – 22% when converting energy. It occupies cabling, inverters, connections, etc.

> **Step 5:** *Calculate the number of solar panels needed for this system:* Consider a well-designed 86% efficiency (14% loss) solar system and divide the solar system size (AC) in *step 4* by 0.86. It looks like this: 7.55 kW / 0.86 = 8.78 kW. Suppose you want to use a solar module with a nominal power of 220 watts for the nameplate. In this case, what you need: 8.78 kW × 1000 / 220 W = 39.90 panel. Always round up this number. In this case, 40, 220-watt solar modules are needed to cover 100% of the energy demand.

Sizing Your Standalone Systems: Islanded or off-the-grid PV systems are different from grid-connected inverters. A standalone PV system can be thought of as a kind of banking system. The battery is a bank account. Photovoltaic generators generate energy (yield), charge batteries (deposits), and power consumers consume energy (withdrawals).

The sizing goal of a standalone PV system is an important balance between energy supply and demand. The generalized schematic for sizing of a standalone PV system with energy storage is shown in Figure 1.9. This includes the following important steps:

- Determine the average daily load requirement for each month.
- Perform critical design analysis to determine the month with the highest load-to-solar insolation ratio.
- Size the system battery bank as per the required voltage and required energy storage capacity.

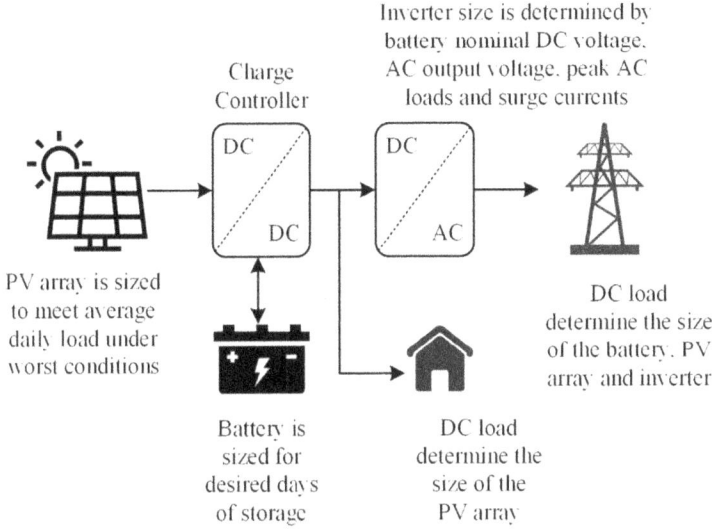

FIGURE 1.9 Generalized schematic for sizing of standalone PV systems with energy storage.

- The PV sizing should meet the average daily load requirements during the lightest and most loaded hours (usually winter).
- The sizing of a standalone PV system begins with the dimensions of the battery and PV array to determine the electrical load followed by the average daily load of the critical design month.

The details for different components of the standalone system are as follows.

1.2.6 System Sizing

PV system sizing for standalone PV systems allows PV system developers or users to accurately size the system based on their expected needs, goals, and budget. It involves a five-step process to do so. These steps include estimating the electrical load, inverter sizing and specifications, battery sizing and specification, array sizing and specifications, and specifying controllers.

Estimating the Electric Load: The first task in PV system design is to determine the system load. Determining the load is easy. Make a list of electrical equipment and or loads supplied by the PV system. Once you have the wattage, a load sizing worksheet can be prepared. Power demand is calculated by multiplying the number of hours per day that a particular device operates each day. For existing buildings, another way is to get the consumption value from the electricity bill. It shows the actual usage for 12 months.

Important Steps for Load Analysis: The load is determined by listing all devices along with their ratings and uptime and summing them up to get the total average energy requirement for watt-hours or kilowatt-hours. The following worksheet provides ideas on how to estimate the load. It makes sense to list both AC and DC loads separately as the inverter only needs to be sized according to AC requirements. We apply inverter efficiency to determine the DC energy required for the AC load. We also apply adjustment factors to adjust DC and AC loads. This gives you a "tuned wattage." Next, the "average daily load" is calculated by multiplying the set wattage by the usage time per day.

Calculation and Explanation:

	Parameters	Values	Explanation
A1	Inverter efficiency	0.85 – 0.9	This quantity is used as a power adjustment factor when the current is changed from DC to AC.
A2	Battery bus voltage	$12V$, $24V$ or $48V$	Battery bus voltage corresponds to the required DC input voltage for the inverter. Follow the below guidance. • Use 12V up to 1 kWh • Use 24V from 1 kWh to 3–4 kWh • Use 48V exceeding 4 kWh.
A3	Inverter AC voltage	Various voltage and frequency levels	The output voltage of the inverter is designed for single phase or three phase at different frequencies.
A4	Rated wattage	Watts	The rated wattage is listed for each appliance.
A5	Adjustment factor	1.0 for DC 0.85 for AC	The adjustment factor is related to the efficiency of the inverter. The efficiency of the inverter varies anywhere between 0.85 and 0.9. For DC loads operating straight from the battery bank, an adjustment factor of 1.0 is used.
A6	Adjusted wattage	$\dfrac{A4}{A5}$	Dividing the rated wattage ($A4$) by the adjustment factor ($A5$) adjusts the wattage to compensate for the inverter's inefficiency and gives the actual wattage consumed from the battery bank.
A7	Hours per day used	hours	The number of hours each appliance is used per day. The duty cycle, or actual time of load operation, must be considered here.
A8	Energy per day	$A6 \times A7$	The amount of energy each appliance requires per day is determined by multiplying each appliance's adjusted wattage ($A6$) by the number of hours used per day ($A7$).
A9	Total energy demand per day	Sum of $A8$	The sum of the quantities in column ($A8$) determines the total energy demand required by the appliances per day. It is calculated in watt-hours.
A10	Total amp-hour demand per day	$A9/A2$	The battery storage is sized independently of the photovoltaic array. In order to size the battery bank, the total electrical load is converted from watt-hours to amp-hours.

(Continued)

	Parameters	Values	Explanation
A11	Maximum AC power requirement	Sum of $A4$	This value is the maximum continuous AC power output required of the inverter if all loads were to operate simultaneously. This does not include surge requirements. The peak or surge requirement (due to motor starting, etc.) must also be considered when selecting an inverter.
A12	Maximum DC power requirement	Sum of $A6$	This value ($A12$) is the maximum DC input power required by the inverter and is necessary to determine wire sizes fusing and disconnect requirements. If load management techniques are employed to eliminate the possibility of loads operating simultaneously, the inverter's maximum output requirements may be reduced accordingly.

1.2.7 BATTERY SIZING

The standalone system battery is sized to store the energy generated by the assembly for use at system load as needed. The total battery capacity required depends on the following factors:

- The number of storage days required to cover the system load without recharging from PV.
- Maximum allowable discharge depth.
- Temperature and discharge rate.
- System loss and efficiency.
- The system voltage defines the number of battery cells required in series.
- The total capacity required defines the number of parallel battery strings required.

1.2.7.1 Days of Storage or Autonomy

- Autonomy is the number of days a fully charged battery can withstand system load without being recharged by a PV generator.
- Longer backup times are used to increase the availability of more important applications and systems but at a higher cost due to the larger battery required.

Important: The battery must be able to meet both the power and energy needs of the system. As a rule of thumb, the minimum autonomy with regular load should be 3 days. For critical loads, autonomy should be at least 3 days, depending on the weather conditions in each region.

1.2.7.2   Factors Affecting Battery Sizing

- The autonomy and maximum permissible discharge depth described defining the total amount of battery capacity required for a particular system load.
- The longer the autonomy time, the larger the battery size, the better the availability, and the lower the average daily discharge depth.
- The higher the allowable depth of discharge, the higher the availability of the system but at the expense of battery condition.
- The nominal capacity of a battery is affected by battery temperature, discharge rate, and age.

The worst-case scenario sizing of the battery bank not only allows the PV system to cover the load on the building in all conditions but also increases the likelihood of minimizing the battery's seasonal discharge depth. It is also important to consider the behavior and the importance of your application.

1.2.7.3   Sample Worksheet for Battery Sizing

Item	Parameters	Values	Explanation
B1	Days of storage desired/required (autonomy)	Days	Battery storage systems are typically designed to provide the required electrical energy for 7 days in the absence of solar radiation. This period is considered the average storage level. Less important applications may require 34 days of storage. This will extend battery life and shorten battery life. For critical applications that endanger public safety, storage for several days may be desirable.
B2	Allowable depth of discharge limit (decimal)	0.8	This is the maximum amount of capacity that can be withdrawn from the battery. Please note that the selected battery requires this limit or a higher discharge depth. A typical value for a good new battery is 0.8.
B3	Required battery capacity (amp-hours)	$\dfrac{A10 \times B1}{B2}$	The required battery capacity is determined by first multiplying the total amp-hours per day ($A10$) by the required number of storage days ($B1$) and then dividing by the permissible discharge depth ($B2$).
B4	Amp-hour capacity of selected battery Note-1	Refer note-1 below	Once the required amp-hours have been determined ($B3$), you can select the battery or battery cell based on the manufacturer's information. See Note 1 below.

(Continued)

Item	Parameters	Values	Explanation
B5	Number of batteries in parallel	$B3 / B4$	Parallel connections provide higher capacity by summing up the entire amp-hour (Ah).
B6	Number of batteries in series	$A2$/selected battery voltage	The battery achieves the desired system voltage (operating voltage) by connecting multiple cells in series. Each cell adds its potential to derive the total terminal voltage.
B7	Total Number of Batteries	$B5 \times B6$	Multiply the number of parallel batteries ($B5$) by the number of series battery cells ($B6$).
B8	Total battery amp-hour capacity (amp-hours)	$B5 \times B4$	The total nominal capacity of the selected batteries is determined by multiplying the number of batteries connected in parallel ($B5$) by the amp-hour capacity (Ah) of the selected batteries ($B4$).

Note-1 (Refer Item B4): Once the required amp-hours have been determined ($B3$), you can select the battery or battery cell based on the manufacturer's information. Furthermore, the battery capacity can vary depending on the discharge rate, so you should use an amp-hour capacity that corresponds to the number of days you need to store it.

1.2.8 PV Array Sizing

The size of the solar array is determined by the following parameters:

- PV generators for standalone systems are sized to cover the average daily load of a critical design month.
- Solar insolation received at the site.
- System loss, soiling, and higher operating temperatures are taken into account when estimating array performance.
- PV module characteristics.
- The system voltage determines the number of modules connected in series for each source circuit.
- The power and energy requirements of your system determine the total number of parallel source circuits required.

The array has the lowest ratio of daily sunlight to daily exposure and is sized to meet the average daily insolation requirements for the month or season. The average daily power supply (kWh or amp-hours) provided by a PV module can be determined based on the module's output and daily solar insolation (peak of solar radiation). You can then understand the load and output requirements of a single module and size the array. Higher system availability can be achieved by expanding solar generators and batteries.

1.2.8.1 Sample Worksheet for PV Array

Item	Parameters	Values	Explanation
C1	Total energy demand per day (watt-hours)	$A9$	Total energy requirements per day (watt-hours).
C2	Battery round-trip efficiency	0.70 and 0.85	A factor between 0.70 and 0.85 is used to estimate the runout efficiency of the battery. If the battery you choose is relatively efficient and a significant percentage of your energy is consumed by sunlight, use 0.85.
C3	Required array output per day (watt-hours)	$C1/C2$	Dividing the total daily energy demand ($C1$) by the battery reciprocating efficiency ($C2$) determines the array power required per day. Since battery usage is less than 100%, the watt-hour required for the load is adjusted (upward).
C4	Selected PV module max power voltage at STC (Volts)	$P_{max} \times 0.85$	Selected maximum power voltage for PV modules at STC $\times 0.85$. The maximum power voltage is derived from the manufacturer's specifications for the selected PV module and is multiplied by 0.85 to determine the design operating voltage for each module (not the array).
C5	Selected PV module guaranteed power output at STC (watts)	Manufacturer's datasheet	The selected PV module guarantees power output at STC. This number is also taken from the manufacturer information for the selected module.
C6	Peek sum hours at design tilt for design month	hours	Peak hours of sunshine at the optimal slope. It is obtained from the solar radiation data at the design location and the slope of the average day array for the worst month of the year.
C7	Energy output per module per day (watt-hours)	$C5 \times C6$	The amount of energy generated per day by the PV array during the worst month is determined by multiplying the PV output selected at STC ($C5$) by the maximum number of hours of sunshine at the slope calculated.
C8	Module energy output at operating temperature (watt-hours)	$DF \times C7$	The derating factor (DF) multiplied by the power output module ($C7$) gives the average power output of one module. $DF = 0.80$ is suitable for hot climates and critical applications. $DF = 0.90$ is for moderate climates and non-critical applications.

(Continued)

Item	Parameters	Values	Explanation
C9	Number of modules required to meet energy requirements	$C3/C8$	The number of modules required to meet the energy requirements can be determined by dividing the power required per day ($C3$) by the output power of the module at the operating temperature ($C8$).
C10	Number of modules required per string rounded to the next higher integer	$A2/C4$	The number of modules required for each string can be determined by dividing the battery bus voltage ($A2$) by the calculated operating voltage ($C4$) of the modules and then rounding to the next higher integer.
C11	Number of strings in parallel rounded to the next higher integer	$C9/C10$	The number of parallel strings can be determined by dividing the number of modules required to meet the energy requirement ($C9$) by the number of modules required for each string ($C10$) and then rounding that number to the next larger integer.
C12	Number of modules to be purchased	$C10 \times C11$	Multiply the number of modules required for each string ($C10$) by the number of parallel strings ($C11$) to get the number of modules to procure.
C13	Nominal rated PV module output (watts)	Manufacturer data	Nominal power (in watts) of the module specified by the manufacturer. PV modules are usually priced according to the module's nominal power output.
C14	Nominal rated array output (watts)	$C12 \times C13$	Multiply the number of modules you purchase ($C12$) by the module's nominal power rating ($C13$) to determine the array's nominal power rating. This number is used to determine the cost of the PV system.

Note: Because it is easy to make meaningful comparisons of the output of PV modules, PV generator design methods often use current (amperes) rather than power (watts) to account for load requirements. For example, instead of comparing 50 –"watt" modules that may have different operating points, if you specify a PV module that produces 30 "amps" at 12 V at the specified operating temperature, you will get performance, physical size, and cost. Very convenient to compare.

1.2.9 Selecting an Inverter

An inverter is required to convert direct current to alternating current. Standalone inverters are usually voltage-specific, i.e., the inverter must have the same nominal voltage as the battery. Inverters are specified in watts. The input power of the inverter must not be less than the total wattage of the device. Inverter Capacity > $A12$ W.

Important: The size of the islanded system's inverter is measured by a maximum continuous output in watts, and this nominal output must be greater than the total wattage of all connected AC loads. Electrical equipment such as washing machines, dryers, and refrigerators also use electric motors and require a lot of power to start. This high starting power consumption can be more than double the normal power consumption, so the input power of the inverter should ideally be 25% – 30% higher than the rated power of the device.

1.2.10 SIZING THE CONTROLLER

The function of the charge controller is to regulate the charge flowing from the solar panel array to the battery bank to prevent nighttime overcharging and reverse current. The most commonly used charging controllers are pulse width modulation (PWM) or maximum power point tracking (MPPT). The voltage at which a PV module can generate maximum power is called the maximum power point (or peak power voltage). The maximum output depends on the amount of solar radiation, ambient temperature, and solar cell temperature. A typical PV module produces about 17 V of power with a maximum grid voltage measured at a cell temperature of $25°C$. It can drop to about 15 V on very hot days and rise to 18 V on very cold days. The MPPT solar charging controller automatically and efficiently corrects the voltage when it detects fluctuations in the voltage characteristics of the solar cell. To draw the maximum power available, operate the PV module at a voltage close to the point of maximum power. The MPPT solar charging controller allows users to use the PV module with an output voltage higher than the operating voltage of the battery system. For example, if the PV module needs to be located far from the charging controller or battery, the cross-sectional area of the cable should be very large to reduce the voltage drop. The MPPT solar charging controller allows the user to wire the PV module to 24V or 48 V (depending on the charging controller and PV module) to power the 12 V or 24 V battery system. This means that the required cable size is reduced while maintaining the full power of the PV module. The input power of the charging controller is equal to the product of the short-circuit current of the PV module, the number of PV modules connected in parallel, and the factor of safety, with a factor of safety of 1.25.

$$I_{Rated} = \left(N_{PV-parallel} \times I_{sc}\right) \times 1.25 \tag{1.5}$$

where I_{Rated} is the solar charge controller rating, I_{sc} is the short-circuit current, $N_{PV-parallel}$ indicates the number of PV modules in parallel, and 1.25 is the safety factor.

Cable Sizing

The purpose of this step is to estimate the size and type of cable for the following connections:

- Cable between PV modules and batteries.
- Cable between the battery bank and the inverter.
- Cable between the inverter and load.

The equation below can be used to determine the cross section of copper wire.

$$A = \frac{p \times L \times I \times 2}{V_d} \tag{1.6}$$

where p is the resistivity of wire (for copper $p = 1.724 \times 10^{-8}\ \Omega \cdot m$), L is the length of wire (in m), A is the cross-sectional area of cable in mm^2, I is the rated current of regulator in amps, and V_d is the voltage drop in volts.

In both AC and DC wiring, the voltage drop is taken not to exceed 4% value, i.e., $V_d = 4100 \times V$. Typically, voltage V is defined by

- Cable between PV modules and batteries $= 12\ V$, $24\ V$, or $48\ V$
- Cable between the battery bank and the inverter $= 12\ V$, $24\ V$, or $48\ V$
- Cable between the inverter and load $= 110\ V$.

1.3   IRRADIANCE FORECASTING

For both private and commercial applications of the PV panel, the process of irradiance forecasting is necessary for site selection. The identification of the site and position of the panel is crucial for designing the PV system as it determines the power generated and streamlines the component selection in other stages. Few of the concepts that needed to be considered for irradiance assessment are as follows:

- Shade analysis: Shading of the solar panel is one of the most prominent issues as it reduces solar power generation. Many factors can contribute to the issue of panel shading such as
  - Shade from a building or nearby tree
  - Shade from the adjacent solar panel
  - Cloudy weather
  - Shadow from the inter-row for the panels

  Shadow analysis is crucial for finalizing the photovoltaic plant installation. For the rooftop panel, the area for installation is determined on the basis of the shading analysis. Most of the installations are connected in the string to meet the voltage requirement. The shading in a particular section can result in efficiency reduction of the panel. The panel efficiency is directly impacted by the area covered by the shadow, and the area which is in the non-shaded region attempts to compensate for the loss which results in overheating of the panels. As the flow of solar energy is blocked by the shadow, it can cause overheating and result in complete panel damage as well. Shadow analysis is difficult because the pattern not only changes throughout the day but also changes throughout the year because of the angle of elevation of the sun and the azimuth angle which are responsible for defining the angle and position of the shadow. It is necessary to plan accordingly to reduce the influence of the shadow on the efficiency of the panel and attain the maximum outcome from the panel.

- **Sun hours:** The time during which the sunlight is available is referred to as the sun hours, but all time is not productive for solar power generation. The PSHs are defined as 1000 W/m^2 of solar irradiance for 1 hour. This amount of peak is expected in the middle of the day when the sun is at its peak and the panel faces toward the sun. The sun is not at its peak throughout the day. During the early morning and in the evening, the irradiance is about 500 W/m^2 or less, whereas during midday and on a clear sunny day, the irradiance can be anywhere around 1000W/m^2 or even higher. In the case of 500 W/m^2, the sunlight can be represented as 0.5 PSH, whereas for 1100 W/m^2, the representation of peaks hours is 1.1. Figure 1.10 illustrates the variation in irradiance throughout time.

  More than four PSHs are required in a day to consider a particular location good for solar energy generation. However, the incentives for green energy are high, and different governments around the globe are supporting the switch; hence, solar energy will pay back for itself sooner and in long term even help in achieving net-carbon neutrality. In Figure 1.11, the annual sun peak hours of few major cities are compared for different months.

- **Tilt angle:** The tilt angle of the solar panel depends on which part of the world the installation takes place. The maximum output is obtained for the panel when it faces the sun directly. The sun moves across, and its location is higher or lower depending upon the time of the day and seasons. Hence, there is no ideal angle, and the optimum angle of the panel needs to be calculated, which depends upon the location and time of the year as illustrated in Figure 1.12. The angle can be calculated as follows:

  Method 1: The optimum tilt angle can be calculated by adding 15° to the latitude in summer and subtracting 15° for winter.

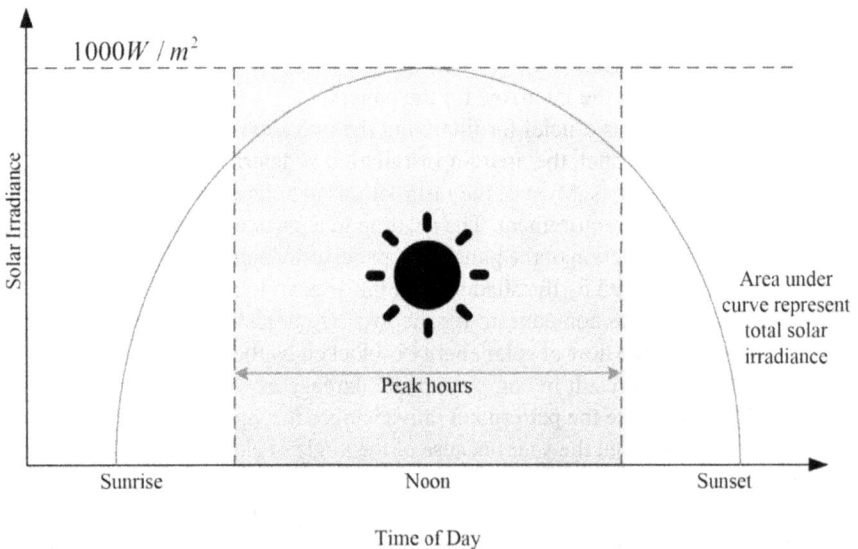

FIGURE 1.10   Solar irradiance curve.

FIGURE 1.11   Annual PSH.

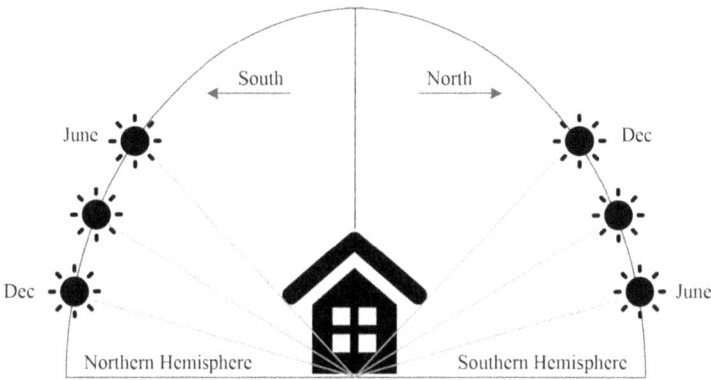

FIGURE 1.12   Tilt angle of the sun.

Method 2: The optimal angle is obtained by multiplying 0.9 to the latitude and adding 29° for winter and, for summer, multiplying 0.9 to the latitude and subtracting 23.5°.

Example 1.4

Calculate the panel tilt angle using methods 1 and 2 when the location of installation is situated at a latitude of 35°?

Solution:

Method 1:
    Summer optimum tilt: $35 - 15 = 20°$
    Winter optimum tilt: $35 + 15 = 50°$
Method 2:
    Summer optimum tilt: $(35 \times 0.9) - 23.5 = 8°$
    Winter optimum tilt: $(35 \times 0.9) + 29 = 60.5°$

1.4   LOAD FORECASTING

The electrical load demand needs to be satisfied by the electrical generation for optimal power system operation. As the battery infrastructure for the storage is not substantial, a generation unit needs to be forecasted so that the electrical infrastructure can be utilized more efficiently. Distribution companies use load-forecasting techniques to predict the load requirement and make sure that the generation unit is ready to satisfy the required demand sufficiently. Transmission companies can use load forecasting to optimize the power flow on the transmission network. Load forecasting can be generally categorized into long-term and short-term load forecasting. Long-term forecasting is used to predict the electrical load for the range of months or years, whereas short-term forecasting is used to predict the load for few hours or days.

The electrical load is majorly influenced, according to Beck et al. [13], by the economy, time, random effect, and weather. The economic factor is influenced by the new installation or development of the infrastructure through the construction of new buildings. All this will impact the load requirement for the generation unit and the infrastructure needed to be updated at the generation end as well. Hence, demand planning and management are necessary to understand the fluctuation and meet the demand, whereas, in the case of time, factors such as seasonal effects, holidays, and weekly cycles impact the electrical load. The seasonal influence is long term on the weather pattern [14]. In summer days, a large cooling load is required in the afternoons and evenings, whereas during winter, a large heating load is to be satisfied during nights and early mornings. The seasonal effect is not only impacted by changes in weather but also in instances such as the daylight-saving time when the load pattern is shifted by an hour. Holidays may further cause fluctuations in the load pattern as they introduce new peaks into the system. Furthermore, during the weekly cycle impact, the fluctuation in load in a week is identified. The periodic variation in load patterns during the start, lunch, and closing times is assisted.

Weather also influences the load pattern significantly. There is a vast impact of weather-sensitive loads such as heating, ventilating, and air-conditioning (HVAC) equipment on the load demand. With the increase in the load of the power system, the load diversity is enhanced, which dampens the effect on the load cycle and smoothens the electrical load profile. Factors such as humidity, solar irradiance, precipitation, and wind speed impact the load profile. For a highly humid day, the cooling unit will run longer to remove excess moisture from the air, whereas for long-irradiance days, heat will radiate in the building interior which will force the cooling equipment to run longer. Precipitation tends to reduce the cooling requirement and leads to low load requirements. Another factor that impacts the load profile is the unplanned system shutdown or system outage.

Few of the commonly used load-forecasting techniques are stochastic time series, state space method, general exponential smoothing, multiple linear regression, artificial neural networks (ANNs), and the knowledge-based expert approach. The forecasting solution for the stochastic time series, state space method, general exponential smoothing, and multiple linear regression is attained by statistical means. In ANNs, it is attained by the historical load and weather data to predict the future load pattern. The input data set is used to train the ANN for approximating the target data set. Furthermore, in the knowledge-based expert approach, rules based on a reference day are used for reshaping the electrical load curve. Few of the load-forecasting methods are explained in Sections 1.4.1–1.4.5.

1.4.1 Multiple Linear Regression (MLR)

The MLR method is used to forecast the weather for a time (t) considering the weather and non-weather variables influence on the electrical load. The variable is identified according to their correlation with the load. Least-square estimation techniques are used to estimate the regression coefficient which is multiplied by the variable [15–17].

The load model for MLR can be expressed as follows:

$$y(t) = a_o + a_1 x_1(t) + \ldots + a_n x_n(t) + a(t)$$

where $y(t)$ represents the electrical load and the explanatory variable in correlation with $y(t)$ is expressed by $x_1(t)\ldots x_n(t)$. The random variable with zero mean and constant variance is denoted by $a(t)$, whereas $a_o, a_1 \ldots, a_n$ represents the regression coefficient.

1.4.2 Stochastic Time Series (STS)

In this method, the historical data is used to predict the future load condition [18–20]. The autoregressive (AR) process forecasts the load $y(t)$ based on the random noise signal and historical load data [21].

$$y(t) = \phi_1 y(t-1) + \phi_2 y(t-2) + \ldots + \phi_p y(t-p) + a(t)$$

In moving average (MA), the load forecasting is done in the form of current and previous random noise signal. The previous forecasting error is used to reconstruct the signal [22].

$$y(t) = a(t) - \Theta_1 a(t-1) - \Theta_2 a(t-2) - \ldots - \Theta_{q1} a(t-q)$$

Furthermore, the autoregressive MA (ARMA) method uses both autoregression (AR) and MA for load forecasting [21,22].

$$y(t) = \phi_1 y(t-1) + \phi_2 y(t-2) + \ldots + \phi_p y(t-p) + a(t) - \Theta_1 a(t-1)$$
$$- \Theta_2 a(t-2) - \ldots - \Theta_{q1} a(t-q)$$

The ARMA can be further modified into autoregressive integrated MA (ARIMA) by performing forecasting in time series along with non-stationary mean. The time series can be in the form of hourly, weekly, monthly, yearly, or other periodicities. The transfer function of one of the modes is illustrated in Figure 1.13.

where $x(t)$ represents the input series variable, whereas $n(t)$ is the colored noise, which is the model from the white noise input.

1.4.3 GENERAL EXPONENTIAL SMOOTHING (GES)

In the GES method of load forecasting, the observed load is represented as follows [23–26]:

$$y(t) = \beta(t)^T f(t) + \varepsilon(t)$$

where $\beta(t)$ is the locally constant coefficient vector and $f(t)$ denotes a vector for stationary fitting and linear independent function. The white noise is represented by $\varepsilon(t)$, whereas T denotes the transpose operation. To represent the forecasted load into the linear combination function, a fitting function is used. The future value of $\beta(t)$ can be estimated for load forecasting with the available fitting function.

$$y(N+l) = f^T(l)\beta(N)[[\text{Tab}]]$$

where l is the lead time represented in hours and the N denotes the last hour when the load is known.

The weighted least-square criterion is used to estimate the coefficient vector $\beta(N)$:

$$\sum_{j=0}^{N-1} w^j \left[y(N-j) - f^T(-j)\beta \right]^2 \quad 0 < w < 1$$

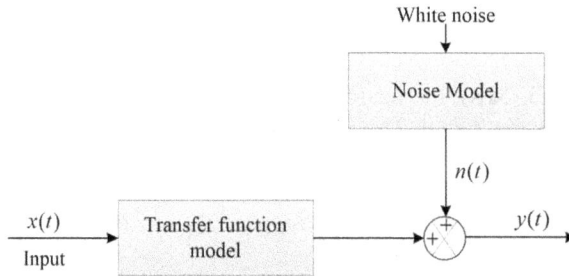

FIGURE 1.13 Transfer function model [21,22].

Hence,

$$\beta(N) = F^{-1}(N)h(N)$$

where

$$F(N) = \sum_{j=0}^{N-1} w^j f(-j) f^T(-j)[[\text{Tab}]]$$

$$h(N) = \sum_{j=0}^{N-1} w^j f(-j) y(N-j)[[\text{Tab}]]$$

On updating the coefficient vector, the forecasting is made using:

$$\beta(N+1) = L^T \beta(N) + F^{-1} f(0)[y(N+1) - y(N)][[\text{Tab}]]$$

$$y(N+1+l) = f^T(l)\beta(N+1)$$

where

$$F = \lim_{N\to\infty} F(N)[[\text{Tab}]]$$

and transition matrix (L) satisfies

$$f(t) = Lf(t-1)[[\text{Tab}]]$$

1.4.4 STATE SPACE (SS)

The ARMA forecasting model can be modified into the state space (SS) model. The SS-based model provides more structure compared to the ARMA model. Furthermore, the SS model provides manipulation and concise presentation capability. Multiple variables are present in the case of the SS forecasting model which are used to model the SS formulation as a state variable. The basing formulas for state space formulation are:

SS Equations: $X(k+1) = \phi(k)X(k) + W(k)$
Measurement Equation: $Z(k) = H(k)X(k) + V(k)[[\text{Tab}]]$

where $X(k)$ represents the $(n \times 1)$ state vector process for the t_k instances and the $\phi(k)$ denotes $(n \times n)$ state transition matrix in relation to $X(k)$ to $X(k+1)$ in the absence of the forcing function, whereas white noise $(n \times 1)$ with covariance $Q(k)$ is represented by $W(k)$. The load measurement for the matrix $(m \times 1)$ at t_k instance is denoted by $Z(k)$, whereas $H(k)$ represents the $(m \times n)$ matrix related to $X(k)$ to $Z(k)$, and the load measurement error matrix $(m \times 1)$ which consists of white noise is represented by $V(k)$.

The SS-based forecasting is not as common as ARMA, as for the ARMA model of forecasting, fewer variables and parameters are required, and the estimation of $R(k)$ and $Q(k)$ is difficult [27–29].

1.4.5   KNOWLEDGE-BASED SYSTEMS (KBSs)

The KBS is a system that emulates the decision-making capability of the line operators when load forecasting is available. During the operation of KBSs, logical decision-making is converted into the form of programming. Data such as historical electrical data and weather data are used to emulate the output decision [30–33]. The KBS programmer explores the relation between weather, time, season, and load for the construction of knowledge-based decisions. The knowledge-based system can be updated based on new data, and the decision process can be updated based on new information and amendments in grid standards.

1.4.6   ARTIFICIAL NEURAL NETWORK (ANN)

The ANN is known as a decision-making mathematical analysis tool that simulates the process of the human brain. The ANN can be trained based on the load data and parameters such as weather and time details [34–37]. Once the training process is completed, the algorithm is tested with predicted data input. Predicted data can be a lagging load [34–37]. The forecasted output is compared with the actual output to identify the error in forecasting. The forecasting error can be present in the form of root-mean-square error or mean absolute percentage error [38–40].

$$\text{Root-mean-square error} = \sqrt{\frac{1}{N} \times \sum_{t=1}^{N} \left(y_t - \widehat{y}_t \right)^2}$$

$$\text{Mean absolute percentage error} = \frac{1}{N} \times \sum_{t=1}^{N} \frac{\left| y_t - \widehat{y}_t \right|}{y_t} \times 100$$

where the number of samples is represented by N, whereas the actual and forecasted load for time t is denoted by y_t and $\hat{y}_t$, respectively.

The ANN architecture consists of an input layer where the data are received and further processed in the hidden layer. The processed data are passed to the rule layer for decision selection based on the input data. The data are further processed before providing the final output decision. The ANN architecture is represented in Figure 1.14.

An issue of overfitting can be created by the ANN backpropagation techniques. The overfitting is created due to network complexity and overtraining. This type of ANN is not capable to establish a relationship between input and output data. Cross-validation can be performed for avoiding the overtraining issue. The data are divided into training and validation data sets [41,42]. The ANN is trained based on the train data set, and the validation data set is used to test the training data after a few iterations of training [43].

A brief comparison between different forecasting techniques is presented in Table 1.2.

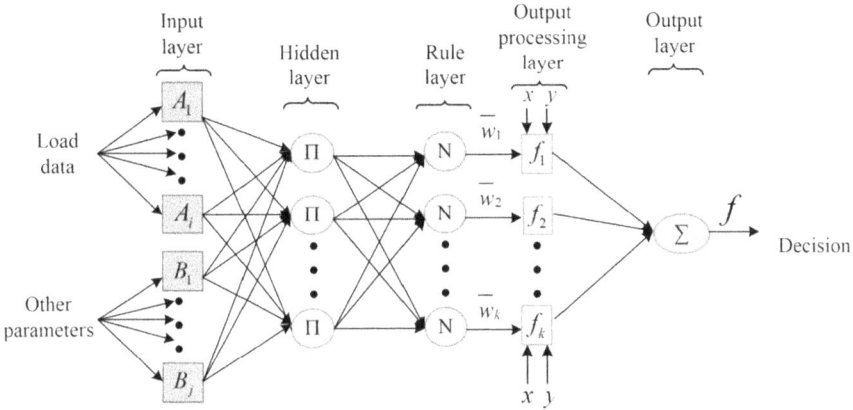

FIGURE 1.14 Generalized architecture of an ANN.

TABLE 1.2
COMPARISON BETWEEN DIFFERENT FORECASTING TECHNIQUES

Method	Advantage and Disadvantage	Usability
Stochastic time series	Simple to implement and interpret but highly sensitive.	*****
State space method	A lot of variables are used to determine the forecasting; however, the estimation of a few of those parameters is very difficult.	*
GES		*
MLR		***
Artificial neural network	Modeling is able to relate different data present in the network with each other. The method is highly sensitive to the dimensional data.	****
Knowledge-based approach		**

***** = Very High, **** = High, *** = Moderate, ** = Low, * = Very Low.

1.5 SUMMARY

This chapter identifies the requirements for the development of a PV system in a defined area and location. A case study is analyzed on how the location of the panel, the capacity to be installed, irradiance received during the day at the site, etc. are channeled during the installation of PV systems and their grid integration around the world. Despite the rapid increase in the installation of PV systems and their impact on covering the very large-scale load demand, there are many technical and economic challenges that are associated with them. These challenges stress the utility grid during the transition between off-generation and high peak hours. Furthermore, the uncertainty in energy generation, use of energy storage systems, requirement of additional infrastructure, monitoring equipment, protection and security, and high-level control algorithms are considered to burden the system operation in terms of both economic and technical points of view.

1.6 QUESTIONS

1. What is a grid?
2. What is an inverter?
3. What is a grid-connected inverter?
4. What is the global energy generation scenario?
5. What is solar insolation? How it is important for solar PV operation.
6. How is PV array sizing done in solar PV plants?
7. What is MPPT? How it is implemented?
8. What is the PSH?
9. What is the importance of forecasting in solar PV-based design?
10. What techniques are used in load forecasting?
11. What techniques are used in solar irradiance forecasting?
12. What is the tilt angle?
13. What is shade analysis?
14. What are the criteria to choose MPPT for a solar PV plant?
15. What are the criteria for selecting a solar inverter?
16. How solar inverter is different from a normal inverter?
17. What is short-circuit current?
18. What is open-circuit voltage?
19. What is MPP?
20. What is the FF?
21. On a clear sunny day, the 4 kWp PV system received 8 peak hours of sunshine hours. The total loss (DF) of the system is estimated to be 0.6 (60%). Estimate the expected output of the system.
22. On a clear sunny day, the 3 kWp PV system received 4 PSHs. Estimate the energy yield from the system.

REFERENCES

[1] International Energy Agnecy, "Global energy review 2020," *Glob. Energy Rev. 2020*, 2020, doi: 10.1787/a60abbf2-en.
[2] Ingram, Elizabeth. "World Adds Record New Renewable Energy Capacity in 2020." Renewable Energy World, 2021. https://www.renewableenergyworld.com/baseload/world-adds-record-new-renewable-energy-capacity-in-2020/#gref (accessed Dec. 21, 2022).
[3] bp p.l.c. "BP Statistical Review of World Energy 2022, (71st Edition)." [Online] London: BP Statistical Review of World Energy, 2022. https://www.bp.com/content/dam/bp/business-sites/en/global/corporate/pdfs/energy-economics/statistical-review/bp-stats-review-2022-full-report.pdf (accessed Dec. 21, 2022).

[4] M. Larsson, *Global Energy Transformation: Four Necessary Steps to Make Clean Energy the Next Success Story*, Basingstoke, Palgrave Macmillian, 2009.

[5] S. P. Deolalkar, "Solar power," in *Designing Green Cement Plants*, Elsevier, 2016, pp. 251–258.

[6] Z. Usman, J. Tah, H. Abanda, and C. Nche, "A critical appraisal of PV-systems' performance," *Buildings*, vol. 10, no. 11, p. 192, Oct. 2020, doi: 10.3390/buildings10110192.

[7] K. N. Nwaigwe, P. Mutabilwa, and E. Dintwa, "An overview of solar power (PV systems) integration into electricity grids," *Mater. Sci. Energy Technol.*, vol. 2, no. 3, pp. 629–633, Dec. 2019, doi: 10.1016/j.mset.2019.07.002.

[8] M. A. Al Mamun, M. Hasanuzzaman, and J. Selvaraj, "Experimental investigation of the effect of partial shading on photovoltaic performance," *IET Renew. Power Gener.*, vol. 11, no. 7, pp. 912–921, Jun. 2017, doi: 10.1049/iet-rpg.2016.0902.

[9] E. Biyik et al., "A key review of building integrated photovoltaic (BIPV) systems," *Eng. Sci. Technol. an Int. J.*, vol. 20, no. 3, pp. 833–858, Jun. 2017, doi: 10.1016/j.jestch.2017.01.009.

[10] C. Perpiña Castillo, F. Batista e Silva, and C. Lavalle, "An assessment of the regional potential for solar power generation in EU-28," *Energy Policy*, vol. 88, pp. 86–99, 2016, doi: 10.1016/j.enpol.2015.10.004.

[11] A. Alshahrani, S. Omer, Y. Su, E. Mohamed, and S. Alotaibi, "The technical challenges facing the integration of small-scale and large-scale PV systems into the grid: A critical review," *Electronics*, vol. 8, no. 12, p. 1443, Dec. 2019, doi: 10.3390/electronics8121443.

[12] A. Tahouri, "*Evaluation of Windows and Energy Performance Case-Study: Colored Building, Faculty of Architecture (EMU) Evaluation of Windows and Energy Performance*," Case-Study: Colored Building, Faculty of Architecture (EMU) Ali Tahouri submitted to the Institute, 2016.

[13] T. Beck, H. Kondziella, G. Huard, and T. Bruckner, "Assessing the influence of the temporal resolution of electrical load and PV generation profiles on self-consumption and sizing of PV-battery systems," *Appl. Energy*, vol. 173, pp. 331–342, 2016, doi: 10.1016/j.apenergy.2016.04.050.

[14] S. G. We, M. One, P. I. D. E. Pid, and P. I. Degradation, "The five most common problems with solar panels," pp. 1–2, 2017. [Online]. Available: https://greensolver.net/en/news/the-five-most-common-problems-with-solar-panels?tmpl=component&print=1&format=print.

[15] A. F. Siegel, "Multiple regression," in *Practical Business Statistics*, Elsevier, 2016, pp. 355–418.

[16] W. N. van Wieringen, "Lecture Notes on Ridge Regression," 2015. [Online]. Available: http://arxiv.org/abs/1509.09169.

[17] M. W. Watkins, "Exploratory factor analysis: a guide to best practice," *J. Black Psychol.*, vol. 44, no. 3, pp. 219–246, Apr. 2018, doi: 10.1177/0095798418771807.

[18] K. A. Agyeman, G. Kim, H. Jo, S. Park, and S. Han, "An ensemble stochastic forecasting framework for variable distributed demand loads," *Energies*, vol. 13, no. 10, p. 2658, May 2020, doi: 10.3390/en13102658.

[19] Y.-F. Tan, L.-Y. Ong, M.-C. Leow, and Y.-X. Goh, "Exploring time-series forecasting models for dynamic pricing in digital signage advertising," *Futur. Internet*, vol. 13, no. 10, p. 241, Sep. 2021, doi: 10.3390/fi13100241.

[20] F. Petropoulos et al., "Forecasting: theory and practice," *Int. J. Forecast.*, Jan. 2022, doi: 10.1016/j.ijforecast.2021.11.001.

[21] M. Zhang, "Time Series : Autoregressive Models," p. 77, 2018.

[22] E. Chodakowska, J. Nazarko, and Ł. Nazarko, "ARIMA models in electrical load forecasting and their robustness to noise," *Energies*, vol. 14, no. 23, p. 7952, Nov. 2021, doi: 10.3390/en14237952.

[23] R. Göb, K. Lurz, and A. Pievatolo, "Electrical load forecasting by exponential smoothing with covariates," *Appl. Stoch. Model. Bus. Ind.*, vol. 29, no. 6, pp. 629–645, Nov. 2013, doi: 10.1002/asmb.2008.

[24] M. Jacob, C. Neves, and D. Vukadinović Greetham, "Short term load forecasting," in *Forecasting and Assessing Risk of Individual Electricity Peaks*, Springer, Berlin, 2020, pp. 15–37.

[25] P. Ji, D. Xiong, P. Wang, and J. Chen, "A study on exponential smoothing model for load forecasting," in *2012 Asia-Pacific Power and Energy Engineering Conference*, Mar. pp. 1–4, 2012, doi: 10.1109/APPEEC.2012.6307555.

[26] J. Mi, L. Fan, X. Duan, and Y. Qiu, "Short-term power load forecasting method based on improved exponential smoothing grey model," *Math. Probl. Eng.*, vol. 2018, pp. 1–11, 2018, doi: 10.1155/2018/3894723.

[27] Hyndman J. Rob, K. B. Anne, S. D. Ralph, and G. Simone, "A state space framework for automatic forecasting using exponential smoothing methods," *Int. J. Forecast.*, vol. 18, no. 3, pp. 439–454, 2002.

[28] R. Adhikari and R.. Agrawal, "An introductory study on time series modeling and forecasting Ratnadip Adhikari R. K. Agrawal," *arXiv Prepr. arXiv1302.6613*, vol. 1302.6613, pp. 1–68, 2013.

[29] M. Elsaraiti and A. Merabet, "A comparative analysis of the arima and lstm predictive models and their effectiveness for predicting wind speed," *Energies*, vol. 14, no. 20, 2021, doi: 10.3390/en14206782.

[30] B. K. Chauhan and P. K. Shukla, "Electric load forecasting using fuzzy knowledge base system with improved accuracy," in *Congress on Intelligent Systems*, Springer, Singapore, pp. 259–273, 2021.

[31] M. C. Falvo, R. Lamedica, S. Pierazzo, and A. Prudenzi, "A knowledge based system for medium term load forecasting," in *Transmission and Distribution Conference and Exhibition, 2005/2006 PES TD*, pp. 1291–1295, doi: 10.1109/TDC.2006.1668697.

[32] M. S. Kandil, S. M. El-Debeiky, and N. E. Hasanien, "Long-term load forecasting for fast-developing utility using a knowledge-based expert system," *IEEE Power Eng. Rev.*, vol. 22, no. 4, pp. 78–78, Apr. 2002, doi: 10.1109/MPER.2002.4312144.

[33] M. R. N. Kalhori, I. T. Emami, F. Fallahi, and M. Tabarzadi, "A data-driven knowledge-based system with reasoning under uncertain evidence for regional long-term hourly load forecasting," *Appl. Energy*, vol. 314, p. 118975, May 2022, doi: 10.1016/j.apenergy.2022.118975.

[34] M. Ding, L. Wang, and R. Bi, "An ANN-based approach for forecasting the power output of photovoltaic system," *Procedia Environ. Sci.*, vol. 11, pp. 1308–1315, 2011, doi: 10.1016/j.proenv.2011.12.196.

[35] Ö. Kişi and E. Uncuoğlu, "Comparison of three back-propagation training algorithms for two case studies," *Indian J. Eng. Mater. Sci.*, vol. 12, no. 5, pp. 434–442, 2005.

[36] A. Nespoli et al., "Day-ahead photovoltaic forecasting: A comparison of the most effective techniques," *Energies*, vol. 12, no. 9, p. 1621, Apr. 2019, doi: 10.3390/en12091621.

[37] M. Radicioni et al., "Power forecasting of a photovoltaic plant located in ENEA Casaccia research center," *Energies*, vol. 14, no. 3, p. 707, Jan. 2021, doi: 10.3390/en14030707.

[38] G. A. Papacharalampous and H. Tyralis, "Evaluation of random forests and prophet for daily streamflow forecasting," *Adv. Geosci.*, vol. 45, pp. 201–208, Aug. 2018, doi: 10.5194/adgeo-45-201-2018.

[39] H. D. Kambezidis, "The solar resource," in *Comprehensive Renewable Energy*, Elsevier, 2012, pp. 27–84.

[40] A. Botchkarev, "Performance Metrics (Error Measures) in Machine Learning Regression, Forecasting and Prognostics: Properties and Typology," pp. 1–37, 2018, [Online]. Available: http://arxiv.org/abs/1809.03006.

[41] T. Shaikhina and N. A. Khovanova, "Handling limited datasets with neural networks in medical applications: A small-data approach," *Artif. Intell. Med.*, vol. 75, pp. 51–63, Jan. 2017, doi: 10.1016/j.artmed.2016.12.003.

[42] S. L. Özesmi, C. O. Tan, and U. Özesmi, "Methodological issues in building, training, and testing artificial neural networks in ecological applications," *Ecol. Modell.*, vol. 195, no. 1–2, pp. 83–93, May 2006, doi: 10.1016/j.ecolmodel.2005.11.012.

[43] J. Barzilai, "On neural-network training algorithms," in *Human-Machine Shared Contexts*, Elsevier, 2020, pp. 307–313.

2 Power Electronics Converter Designing. Part 1
DC Side

2.1 INTRODUCTION

The solar photovoltaic (PV)-based power plant is considered one of the most valuable, abundant, and low maintenance clean renewable energy sources. This solar PV-based system needs a DC–DC converter to control and regulate the varying output of solar panels.

The DC–DC converters are classified mainly as isolated and non-isolated DC–DC converters. Examples of non-isolated DC–DC converters are buck, Boost, Cuk, etc., and examples of isolated converters are flyback, Forward, Push-Pull, etc.

In this chapter, the application of DC–DC converter in solar PV-based plants is discussed along with the operating principle of four main DC–DC converters, i.e., two non-isolated and two isolated converters, which are commonly used with solar PV plants. In addition, a summary of other DC–DC converters along with their mathematical equation is given.

The maximum power point tracker (MPPT) is discussed, which is implemented in the control circuit of DC–DC converters. The calculation of the optimized duty cycle for operating at MPP is also elaborated. The types of MPPT and their limitation are also discussed.

2.2 APPLICATION OF DC–DC CONVERTERS IN SOLAR PV SYSTEMS

A DC–DC converter is mainly used either to step up or to step down the DC voltage at the load end. The DC–DC converter is used in solar photovoltaic (PV)-based power plants and has an important role in it, i.e., stepping up or stepping down the DC voltage at its output terminals and implementing the MPPT through its control circuit by varying the duty cycle. Many new types of DC–DC converters are invented due to the need and advancement in power electronics technologies.

The main application of DC–DC converter in solar PV plants is shown in Figure 2.1.

It is used to regulate the DC output voltage as per the requirement, to implement MPPT in its control circuit, to work as battery chargers, and to derive auxiliary power supply for various DC loads and used in solar PV-based street light applications.

DOI: 10.1201/9781003257189-2

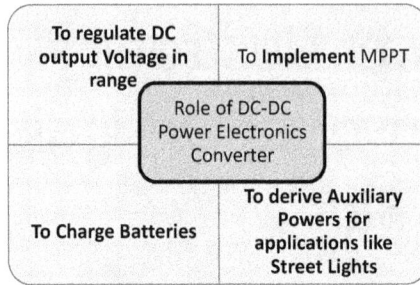

FIGURE 2.1   Main function of DC–DC converters in solar PV plants.

Moreover, the DC–DC converter is connected between the output of solar PV panels and the input of grid-connected and standalone inverters, as shown in Figures 2.2–2.4.

2.3   DC–DC CONVERTER

The DC–DC converter is broadly classified into two categories, i.e., isolated type and non-isolated type DC–DC converters which are widely used in solar PV applications.

The main types of such converters are shown in Figure 2.5.

The major difference between the two types of converters is in cost and galvanic isolation. The isolated type DC–DC has galvanic isolation between the input and the output because of a magnetically coupled element and at the same time, the cost is high. Flyback, Resonant, Forward, Push-pull, and bridge DC–DC converters are examples of isolated converters. The non-isolated type converters are Cuk, single-ended primary inductance converter (SEPIC), Boost, Buck-Boost, etc.

2.3.1   Buck Converter

The buck converter, also known as a step-down converter, is used to step down the input voltage at the output side. The schematic of the converter is shown in Figure 2.6. This converter is a switched-mode power supply containing a diode, a power switch, and an energy storage element, i.e., a capacitor and an inductor. Here, the input voltage source feeds the controllable power switch which is operated with a pulse width modulation (PWM) either through a time base or with a frequency base. Generally, to eliminate the voltage ripple, a combination of capacitor and inductor is used in the converter [1]. The main application of this converter is in the battery charger circuit [4], solar PV pumping system [5], MPPT tracking, etc. [6].

2.3.1.1   Steady-State Operation Analysis

The schematic of the buck DC–DC converter is shown in Figure 2.7, where V_{in} is the input DC voltage (which is equivalent to the voltage from the solar PV panel), S_w is the power switch (whose duty cycle is controlled), D is the diode, and V_o is the output voltage.

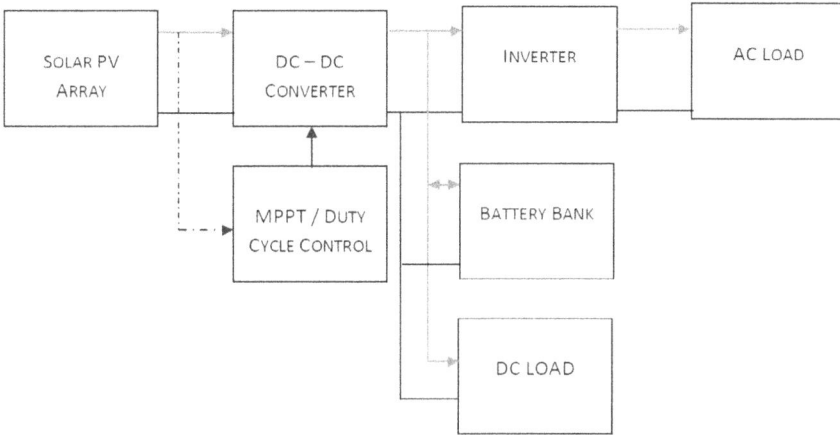

(a): Typical Connection of Standalone Solar PV based Power Plant

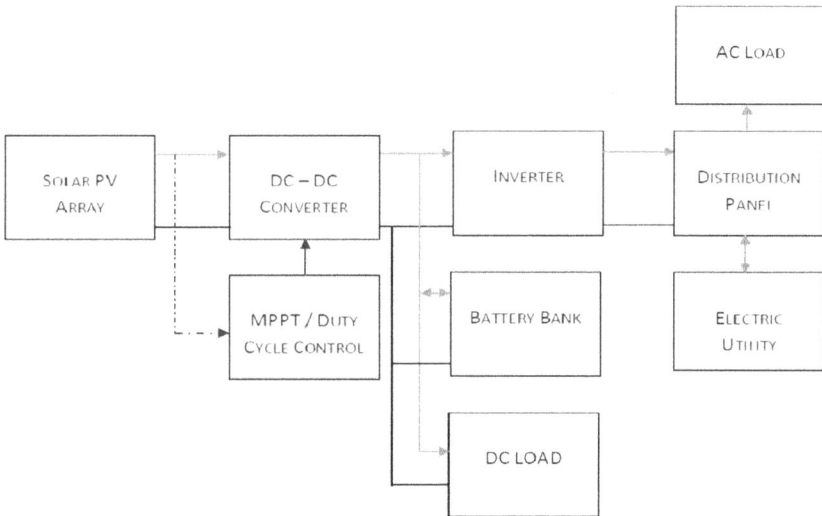

(b): Typical Connection of Grid Connected Solar PV based Power Plant

FIGURE 2.2 (a) Typical connection of standalone solar PV-based power plants. (b) Typical connection of grid-connected solar PV-based power plant.

The operation of the buck DC–DC converter can be divided into two stages, i.e., when the switch S_w is ON and when it is OFF.

The schematic of the buck DC–DC converter when switch S is on and off is shown in Figures 2.8 and 2.9, respectively.

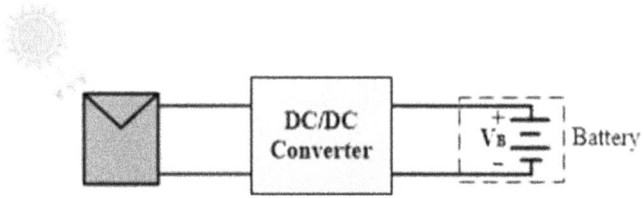

FIGURE 2.3   Battery charging circuit.

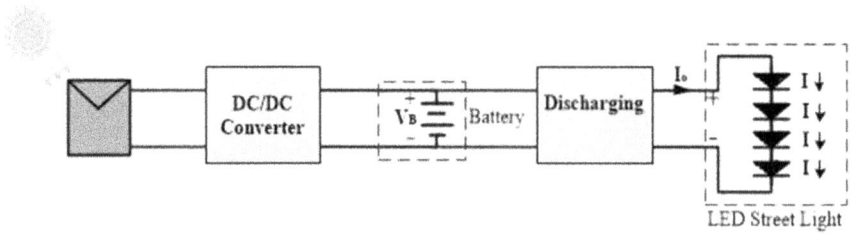

FIGURE 2.4   LED street light circuit.

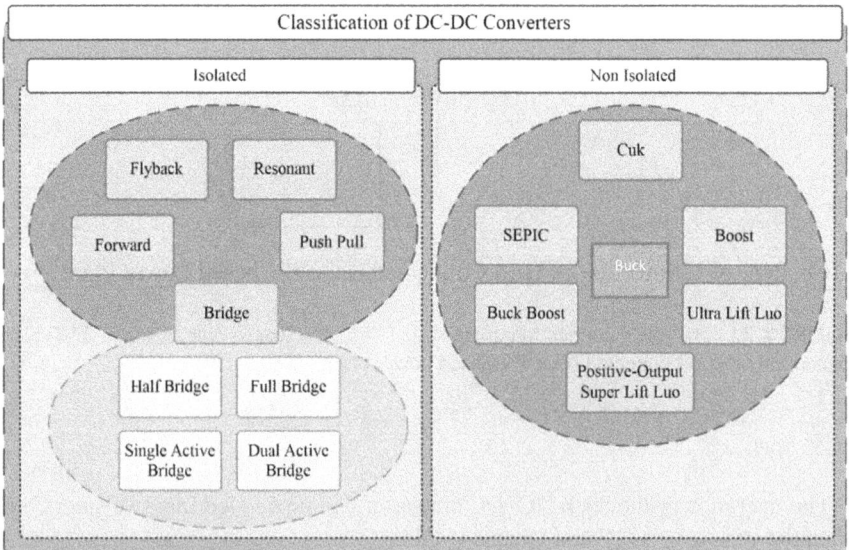

FIGURE 2.5   Classification of DC–DC converters used in solar PV applications.

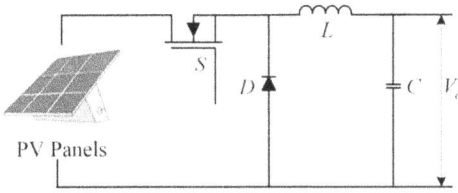

FIGURE 2.6   Schematic of a buck DC–DC converter.

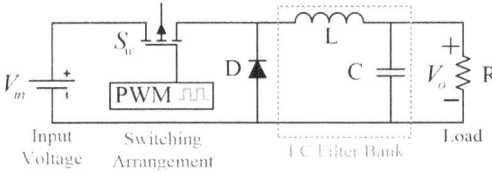

FIGURE 2.7   Schematic of a buck DC–DC converter.

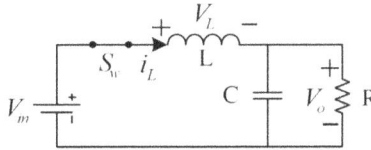

FIGURE 2.8   Schematic of a buck DC–DC converter – when switch S is ON.

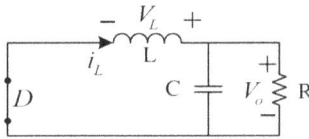

FIGURE 2.9   Schematic of a buck DC–DC converter – when switch S is ON.

The waveforms for the voltage and current are shown in Figure 2.10 for continuous current flow in the inductor L. It is assumed that the current rises and falls linearly. The resistance of switch is very low and is neglected in most of the circuit analysis.

The voltage across inductor L is given as follows:

$$e = L\frac{di}{dt} \tag{2.1}$$

2.3.1.2   Steady-State Analysis When Switch S_w Is ON

The inductor current rises linearly from I_1 to I_2 in time t_1, from Figure 2.10,

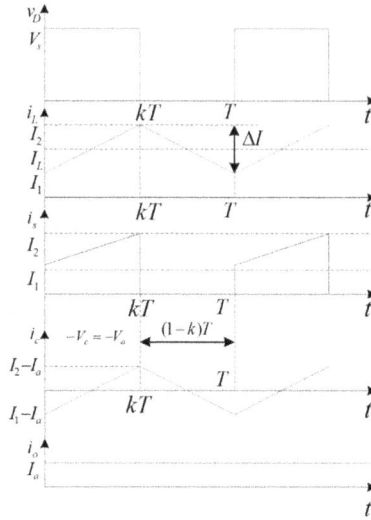

FIGURE 2.10   Waveforms of a buck DC–DC converter (k=d=duty cycle).

$$V_{\text{in}} - V_o = L\frac{I_2 - I_1}{t_1} = L\frac{\Delta I}{t_1} \tag{2.2}$$

$$\Delta I = \frac{(V_{\text{in}} - V_o)t_1}{L} \tag{2.3}$$

2.3.1.3   Steady-State Analysis When Switch S_w Is OFF

The inductor current falls linearly from I_2 to I_1 in time t_2, from Figures 2.9 and 2.10,

$$V_o = \frac{L\Delta I}{t_2} \tag{2.4}$$

$$\Delta I = \frac{V_o t_2}{L} \tag{2.5}$$

$\Delta I = I_2 - I_1$ is the peak-to-peak ripple current of the Inductor L.
Equating equations 2.3 and 2.5,

$$\Delta I = \frac{(V_{\text{in}} - V_o)t_1}{L} = \frac{V_o t_2}{L} \tag{2.6}$$

$$\text{where } t_1 = dT \text{ and } t_2 = (1 - d)T \tag{2.7}$$

d is the duty cycle and T is the time period of switching frequency.

Substituting from equation 2.7 to 2.6:

$$V_o = d\, V_{in} \tag{2.8}$$

Equation 2.8 is the expression of the average output voltage of a buck DC–DC converter.

Since DC–DC converters are very highly efficient circuits, assuming it is a loss-less circuit,

$$V_{in}I_{in} = V_o I_o = d\, V_{in} I_o$$

where I_{in} is the average input current, and I_o is the average output current of the buck converter.

$$I_{in} = d\, I_o \tag{2.9}$$

2.3.1.4 Expression of Peak-to-Peak Inductor Ripple Current

The expression for peak-to-peak inductor ripple current is derived as [2]

$$\Delta I = \frac{V_o\left(V_{in} - V_o\right)}{fLV_{in}} \tag{2.10}$$

$$\Delta I = \frac{V_{in}\, d(1-d)}{fL} \tag{2.11}$$

where f is the switching frequency of switch S_w.

2.3.1.5 Expression of Peak-to-Peak Capacitor Ripple Voltage

The expression for peak-to-peak capacitor ripple voltage is derived as [2]

$$\Delta Vc = \frac{\Delta I}{8fC} \tag{2.12}$$

$$\Delta Vc = \frac{V_o\left(V_{in} - V_o\right)}{8LCf^2\, V_{in}} \tag{2.13}$$

$$\Delta Vc = \frac{V_{in}\, d\,(1-d)}{8\, L\, C\, f^2} \tag{2.14}$$

Example 1 Numerical for Buck DC–DC Regulator

A solar PV-based plant is used to supply power to a resistive load whose value is $R = 50\ \Omega$ and whose voltage across the load is $V_o = 25\ V$. The output voltage of the solar PV panel is 60 V. The peak-to-peak output ripple voltage is 20 mV. The switching frequency is 30 kHz. If the peak-to-peak ripple current of the inductor is limited to 0.8 A, determine:

a. The duty cycle
b. The value of filter inductor L
c. The value of filter capacitor C

Solution:

$V_{in} = 60\ V$
$\Delta Vc = 20\ mV$
$\Delta I = 0.8\ A$
$f = 30\ kHz$
$V_o = 25\ V$

    a. $V_o = d\ V_{in}$
       duty cycle $d = 25/60 = 0.416$

    b. The value of filter inductance L from equation

$$\Delta I = \frac{V_o (V_{in} - V_o)}{fLV_{in}}$$

$$L = 25(60 - 25)/(30000 \times 0.8 \times 60)$$

$$= 25 \times 35 / 1440000 = 607.6\ \mu H$$

    c. The value of filter capacitor C from equation

$$\Delta Vc = \frac{\Delta I}{8fC}$$

$$C = 0.8 / (8 \times 30000 \times 20 \times 10^{-3})$$

$$= 166.67\ \mu F$$

2.3.2 SEPIC CONVERTER

The SEPIC is widely used in DC voltage flickering control and sensor-less control of PV applications. The schematic of the SEPIC is shown in Figure 2.11. While operating the converter, the on-time switching must be larger than the off-time switching to realize a high voltage at the output. Furthermore, this also ensures that the capacitor is fully charged. For any condition, if the switching is not achieved, the

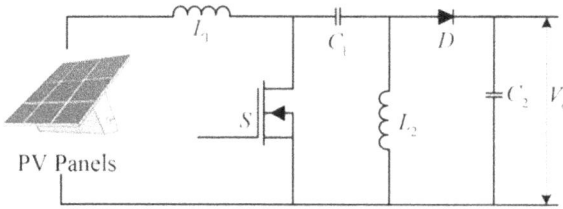

FIGURE 2.11 Schematic of the SEPIC DC–DC converter.

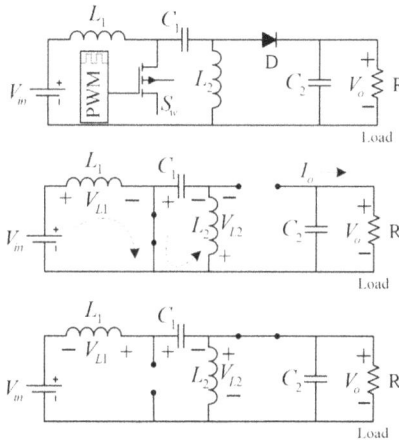

FIGURE 2.12 Operation of the SEPIC DC–DC converter.

converter fails to provide the required output. Besides, the operation of the converter along with a high-frequency transformer achieves an output voltage with minimized ripples. This provides various advantages with key features like continuous output current, minimized output ripples, and minimized switching stress [12,15–17].

Basic switch-mode DC–DC converters suffer from a high current ripple. This creates harmonics. In many applications, these harmonics necessitate using an LC filter. Another issue that can complicate the usage of the buck-boost converter topology is the fact that it inverts the voltage. Cuk converters solve the harmonics problem by using an extra capacitor and inductor. However, both Cuk and buck-boost converters reverse the voltage polarity of the input voltage and cause large amounts of electrical stress on the components, and this can result in device failure or overheating. SEPIC converters solve both of these problems [3].

2.3.2.1 Steady-State Operation Analysis

Applying Kirchhoff's voltage law around the path V_s, L_1, C_1, and L_2 in Figure 2.13. When the switch is open, gives $-V_s + V_{L1} + V_{c1} - V_L = 0$. Using the average of these voltages, the voltage across capacitor C_1 is $V_{c1} = V_s$. When the switch is closed, the voltage across L1 during DT is $V_{L1} = V_s$. When the switch is open, applying KVL

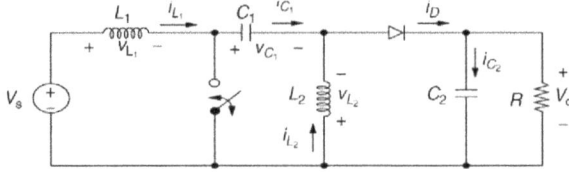

FIGURE 2.13 Schematic of the SEPIC DC–DC converter.

around the outermost path gives $-V_s + V_1 + V_{c1} + V_0 = 0$. Assuming the voltage across C_1 remains constant, then $V_{L1} = -V_0$ for a time period of $(1-D)T$. Using the voltage across the inductor of zero for a periodic operation then $V_s DT - V_0 (1-D)T = 0$, where D is the duty ratio of the switch. Then $V_0 = Vs (D/1-D)$, which is expressed as $D = V_0/(V_0 + V_s)$. This is similar to buck-boost and Cuk converter equations, but with no reverse polarity.

The variation i_{L1} when the switch is closed is $V_{L1} = V_s = L_1(\Delta i_{L1}/DT)$. On Solving for $\Delta i_{L1} = V_s DT/L_1 = V_s D/L_1 f$. For L_2, the average current is determined from Kirchhoff's current law at the node where L_1, C_2, and the diode are connected.

$i_{L2} = i_d - i_{c1}$ and the diode current is $i_D = i_{c2} + I_0$.

The output stage consisting of the diode, C_2, and resistor is the same as the boost converter and so the output voltage ripple is. $\Delta V_0 = V_{c2} = (V_0 D)/(RC_2 f)$. The voltage variation in C_1 is determined from the circuit with the switch closed. Capacitor C_1 current has an average value of I_0 where $\Delta V_{c1} = (\Delta Q_{c1}/C) = (I_0 DT/C)$. Replacing I_0 with V_0/R result in $\Delta V_{c1} = (\Delta Q_{c1}/C) = (V_0 D/RC_1 f)$.

2.3.3 BOOST CONVERTER

The boost converter is referred to as a step-up converter and is used in applications where the voltage magnitude at the output needs to be larger than the input voltage. This converter finds its majority of applications in PV systems to boost the PV voltage [7–9]. The schematic of the boost converter is shown in Figure 2.14. Similar to a buck converter, the boost converter is also a switched-mode power supply containing an inductor, a power switch, a diode, and a capacitor. Here, the input voltage source feeds the inductor which leads to a constant input current. Furthermore, the power switch is operated with a PWM to achieve the required output voltage. Many modifications are available for the basic boost converter in the literature [7–10] to achieve ripple minimization, high voltage gain, and enhanced performance.

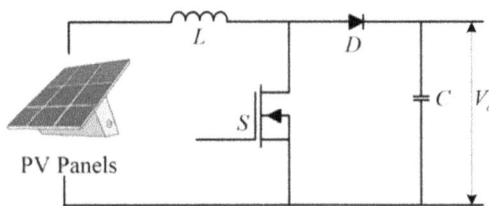

FIGURE 2.14 Schematic of the Boost DC–DC converter.

2.3.4 BUCK-BOOST CONVERTER

The buck-boost converter can operate both in step-down and step-up modes depending upon the duty cycle provided to the converter. The schematic of the converter shown in Figure 2.15 is developed by combining the basic buck converter and boost converter topologies discussed above. This converter found most of its applications in standalone and grid-connected PV systems, and motor drives [11]. Similar to the operation of a buck converter, the input voltage source feeds the controllable power switch in the converter which is operated with a PWM. The literature identified that the continuous current mode operation of the buck-boost converter has lower ripples in the current. Furthermore, a buck-boost converter with two power switches can have the least current and voltage stress on the components operating in the converter. Moreover, to enhance the operation of a basic buck-boost converter, various other topologies like SEPIC [12], Cuk [13], and Luo converters [14] are available in the literature.

2.3.5 CUK CONVERTER

The schematic of a Cuk converter shown in Figure 2.16 has similarities with the basic buck-boost converter except for the fact that the inductor is replaced with a capacitor to achieve the power transfer. This arrangement is also known as a negative-output capacitive energy-based flyback DC–DC converter [18]. Furthermore, the Cuk converter achieves ripple-free output in a system by inverting the output polarity of the converter with suitable connections [13,19,20]. Besides, to improve the efficiency of Cuk converters and achieve optimal bidirectional operation concerning the regulation of voltage and current [21], various modifications are proposed in the literature [22–25]. These studies have identified the application of the converter in various motor drive circuits [18], and renewable energy applications [26–28].

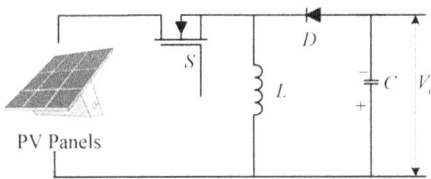

FIGURE 2.15   Schematic of the buck-boost DC–DC converter.

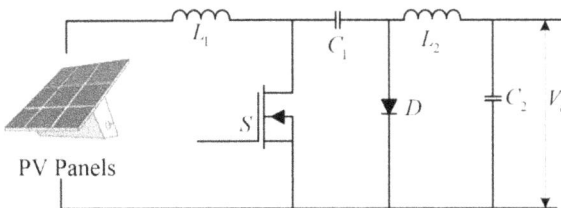

FIGURE 2.16   Schematic of a Cuk DC–DC converter.

2.3.6   POSITIVE-OUTPUT SUPER-LIFT LUO CONVERTER

The positive-output super-lift Luo converter shown in Figure 2.17 was initially introduced by Luo et al. [14] in 2003. This converter is developed with different series
energy storage elements like series inductors and capacitors which provide high output voltage resembling arithmetic progressions. Later, the design of the converter
was modified in reference [29] by adding a high voltage transfer gain, and its operation was enhanced by using a sliding mode controller in reference [30] for achieving
the balance between the voltage regulation and the load current. This converter is
considered to be more powerful when compared to the SEPIC and Cuk converters
discussed above due to its unique features of enhanced efficiency and high output
voltage resembling higher geometric progressions. Moreover, these converters are
still under development for their operation with domestic and industrial PV applications [31,32].

2.3.7   ULTRA-LIFT LUO CONVERTER

The ultra-lift Luo converter is shown in Figure 2.18 [33,34]. This converter combines
the design aspects of voltage and super-lift Luo converters to produce a high voltage
conversion gain. This makes the converter highly efficient among the other non-
isolated DC–DC converters. Furthermore, it is identified that the closed-loop design
of the converter is monotonous as the slightest variation in duty ratio results in large
output voltage variations.

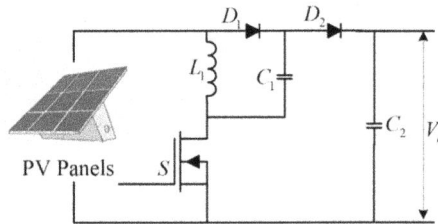

FIGURE 2.17   Schematic of a positive-output super-lift Luo DC–DC converter.

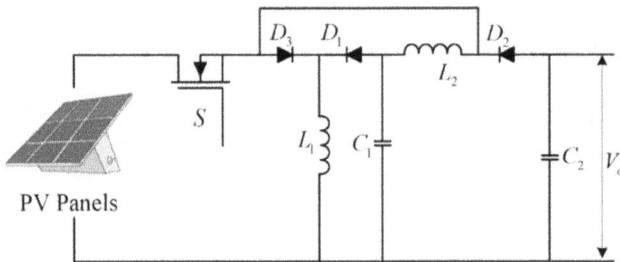

FIGURE 2.18   Schematic of an ultra-lift duo DC–DC converter.

2.3.8 ZETA CONVERTER

The zeta converter combines the advantages of the buck-boost, SEPIC, and Cuk converters. The schematic of the zeta converter is shown in Figure 2.19. When operated in a PV system, the zeta converter enables continuous MPPT over the entire area of the PV curve. Furthermore, the zeta converter provides a non-inverted output voltage that has either an enhanced or diminished value concerning the input voltage [35]. Moreover, to reduce the output ripples and achieve enhanced voltage conversion in continuous and discontinuous modes, new topologies of the zeta converter are developed. These advancements are constituted for operation with the battery storage systems in PV applications.

2.3.9 FLYBACK CONVERTER

The schematic of the flyback DC–DC converter is shown in Figure 2.20. This converter acts as a key solution for higher converter gain requirements by employing transformers in the system. For a transformer with a large air gap to store energy, the flyback converter can be used in high-power applications. Furthermore, the large air gap results in less magnetizing inductance, and the flyback converter provides very less energy transfer efficiency and large leakage flux. Moreover, the flyback converter overcomes the drawback of output polarity inversion and high current flow in the power switch and output diode in Cuk converters. These advantages have seen the application of flyback converters to operate in the discontinuous mode with isolated grid-connected inverters. This application identified unique features like swift dynamic response and less complexity of the converter. Furthermore, the efficiency and decreased ripple content of the converter is enhanced by employing various soft switching techniques.

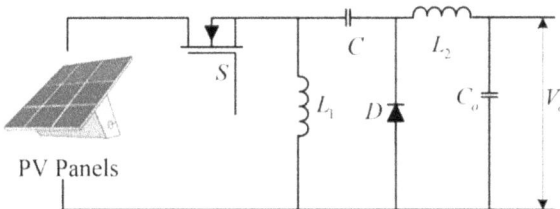

FIGURE 2.19   Schematic of a Zeta DC–DC converter.

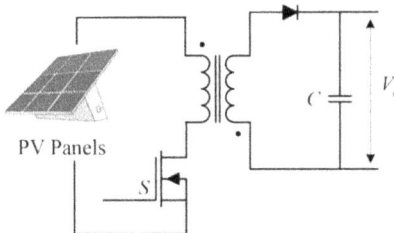

FIGURE 2.20   Schematic of a flyback DC–DC converter.

2.3.10   Three-Port Half-Bridge DC–DC Converter

The schematic of a three-port half-bridge DC–DC converter is shown in Figure 2.21. The converter's primary circuit operates in the buck converter mode with synchronous rectification to provide the high-frequency transformer with a DC bias current. Furthermore, various implementations are projected for post and synchronous regulations to regulate the three ports individually for achieving a single-stage power conversion with modest topology, and simple control. Moreover, to achieve continuous input current with a wide range of zero voltage switching and low ripple, the three switches are operated with an active-clamped half-bridge DC converter. Besides, to achieve a high voltage gain for large input voltage applications, the hybrid secondary rectifier is modified as a dual half-bridge LLC resonant converter. Here, depending on the switching strategy, the hybrid secondary rectifier acts as a quadruple rectifier. The output power and efficiency are improved by employing an interleaved high-performance DC converter. A half-bridge of this topology ensures that the duty cycle of the two interleaved converters is close to 50% for achieving a continuous output current width with less lag and small component size. Furthermore, to reduce electromagnetic interference and voltage stress and achieve high efficiency, the three-level converter is operated with a high-voltage bidirectional half-bridge in high-voltage DC microgrid applications.

2.3.11   Full-Bridge Converter

The full-bridge converter is used to integrate various components of the PV system like the PV array, energy storage device, and the load. The general schematic of a full-bridge converter is shown in Figure 2.22. The arrangement of a full-bridge converter consists of the integration of two buck-boost converters. This is developed to achieve zero voltage switching and single power conversion with the topology. Furthermore, the circulating current losses are minimized by achieving zero voltage switching during the turn-on of switches by operating the full-bridge topology with asymmetrical PWM. Moreover, the use of asymmetrical PWM with a full-bridge converter minimizes the stress on power switches and achieves higher efficiency. Besides, the problem of reverse recovery in the output

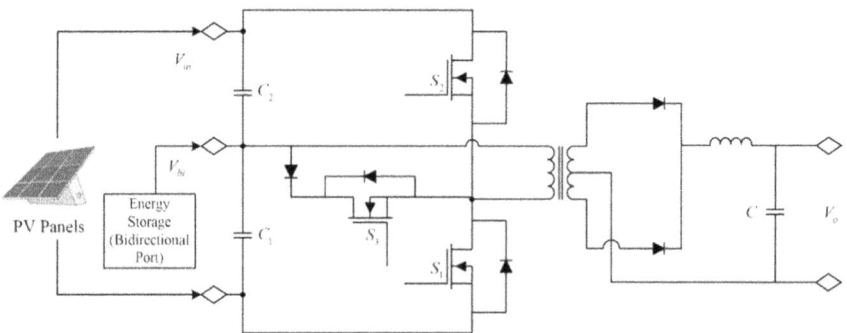

FIGURE 2.21   Schematic of a three-port half-bridge DC–DC converter.

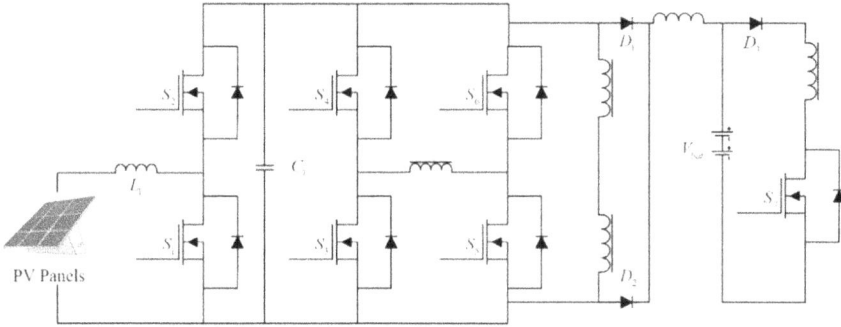

FIGURE 2.22   Schematic of a full-bridge DC–DC converter.

diode is overcome by achieving zero current switching during the switch turn off by combining the resonant part of the circuit with the blocking capacitor and leakage inductance.

2.3.12   DUAL-ACTIVE BRIDGE CONVERTER

The dual-active bridge converter has found its application in standalone hybrid systems due to its advantages with high conversion efficiency, bidirectional power flow, galvanic isolation, and high-power density. The schematic of the converter is shown in Figure 2.23. From the circuit, it can be identified that the high-voltage DC sources feed the primary bridge, and the low-voltage energy storage or load is connected to the secondary bridge. Furthermore, a high-frequency power transformer is used to isolate the two full bridges whose leakage inductance is used as a storage element in the circuit. To enable the bidirectional power flow with the circuit, a square wave is conveniently phase-shifted between both bridges. Moreover, the voltage difference of the storage element is controlled to achieve power conversion with the circuit. The control aspects of the dual-active bridge-isolated bidirectional DC/DC converter are widely monitored through digital controllers. The high-frequency dual-active bridge transformers were developed as an improvement to the existing converters. Besides, an ultra-capacitor-based dual-active bridge converter was developed.

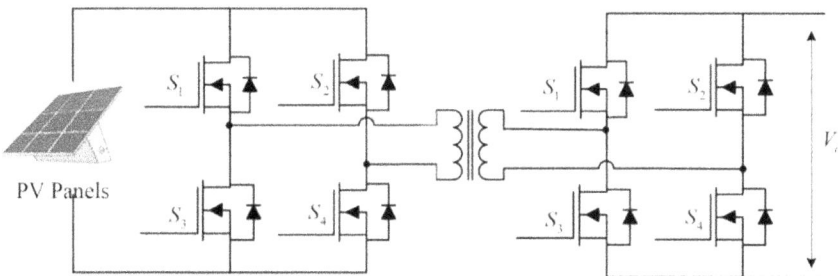

FIGURE 2.23   Schematic of a dual-active bridge DC–DC converter.

2.3.13 MULTI-ELEMENT RESONANT CONVERTER

The multi-element resonant converter is an enhanced topology of the traditional LLC converter. This converter provides the advantages of zero voltage and current switching using a short output circuit, higher efficiencies at the full load operation, and high power density making them adaptable in renewable energy generation applications. The schematic of a three-port multi-element resonant converter is shown in Figure 2.24. The design aspects of this converter involve series and parallel connections of five resonant components in the circuit. These multiple resonant components provide various resonant frequencies in the circuit and their suitable placement helps in transferring the active power of fundamental and third-order harmonics. Furthermore, the parasitic leakage current due to the non-ideal isolated transformer is ignored for this converter. From the literature, it is identified that the zero voltage switching characteristics can be easily achieved for all the power switches in the three ports along with 96% power conversion efficiency.

2.3.14 PUSH-PULL CONVERTER

The push-pull converter also known as a switching converter is shown in Figure 2.25. This converter involves a transformer and operates with the help of a center-tapped primary winding by acting as a forward converter to the transformer core effectively. Furthermore, the push-pull converter has small filters for different available power levels with the circuit. The major advantage with this converter is that the transistor pair in the circuit employs input lines that avail the flow of current through the main winding of the transformer. Besides, the concurrent switching of the transistors draws current from the transformer resulting in a shattered condition for the current at the

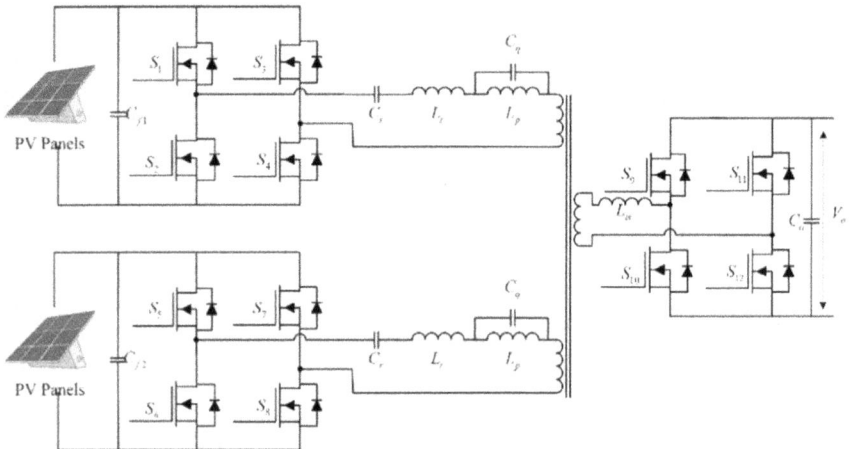

FIGURE 2.24 Schematic of a bidirectional multi-element DC–DC resonant converter.

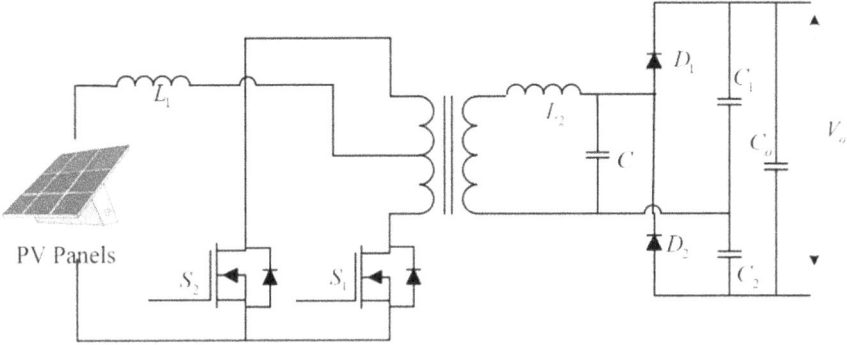

FIGURE 2.25   Schematic of a push-pull DC–DC converter.

TABLE 2.1
Comparison between Different DC–DC Converters

	Converter Type	Optimal Power Demand	Voltage Stress	Efficiency
Non-isolated	Buck	Low	High	Medium
unidirectional	Boost	Low	High	Medium
DC–DC converter	Buck-Boost	Low	Medium	Medium
Non-isolated	Buck-Boost	Low	Medium	Medium
bidirectional	SEPIC	Low	High	Medium
DC–DC converter	Cuk	Low	High	Medium
	Half-bridge	Low	Medium	High
Isolated unidirectional	Flyback	Low	High	High
DC–DC converter	Half-bridge	Low	High	High
	Full bridge	High	Medium	Medium
	Push-pull	High	High	Medium
Isolated bidirectional	Half-bridge	Low	High	Medium
DC–DC converter	Full bridge	High	Low	Medium

line during the switching condition of the half-cycle pair. Moreover, when operated with low input noise, the pull-pull converters have stable input currents and can have efficient high-power applications. Furthermore, the center tapping which only utilizes half of the winding of the transformer at a time resulted in increased copper losses for the circuit.

Furthermore, a summary of different DC–DC converters classified under isolated, non-isolated, unidirectional, and bidirectional are shown in Table 2.1. The summary identifies the important aspects of different converters and compares them with each other (Table 2.2).

TABLE 2.2

Summary of Input Output Voltage Relationship

Buck converter	$D = V_o / V_i$	D = Duty cycle V_o = Output voltage V_i = Input voltage
Boost converter	$\dfrac{1}{1-D} = V_o / V_i$	D = Duty cycle V_o = Output voltage V_i = Input voltage
Buck-boost converter	$\dfrac{D}{1-D} = V_o / V_i$	D = Duty cycle V_o = Output voltage V_i = Input voltage
SEPIC	$\dfrac{D}{1-D} = V_o + V_D / V_i$	V_D = Voltage drops across the diode D = Duty cycle V_o = Output voltage V_i = Input voltage
Cuk converter	$\dfrac{D}{1-D} = \dfrac{V_o}{V_i}$ $\dfrac{D}{1-D} = \dfrac{I_o}{I_i}$	D = Duty cycle V_o = Output voltage V_i = Input voltage I_o = Output current I_i = Input current
Positive-output super-lift Luo converter	$\dfrac{2-D}{1-D} = \dfrac{V_o}{V_i}$	D = Duty cycle V_o = Output voltage V_i = Input voltage
Ultra-lift Luo converter	$\left(\dfrac{j+2-D}{1-D}\right)^n = \dfrac{V_o}{V_i}$	D = Duty cycle n = Number of stages j = Multiple enhanced number V_o = Output voltage V_i = Input voltage
Zeta converter	$\dfrac{D}{1-D} = \dfrac{V_o}{V_i}$ $\dfrac{D}{1-D} = \dfrac{I_o}{I_i}$	D = Duty cycle V_o = Output voltage V_i = Input voltage I_o = Output current I_i = Input current
Flyback converter	$D = \dfrac{V_o'}{V_o' + V_i}$ $V_o' = V_o \dfrac{N_p}{N_s}$	V_o' = Output current I_i = Input current N_s = Secondary winding N_p = Primary winding D = Duty cycle V_o = Output voltage V_i = Input voltage

(Continued)

TABLE 2.2 (Continued)
Summary of Input Output Voltage Relationship

Full-bridge converter	$V_o = \dfrac{N_s}{N_p} V_i D$	N_s = Secondary winding N_p = Primary winding D = Duty cycle V_o = Output voltage V_i = Input voltage
Dual-active bridge converter	$P_o = \dfrac{n V_p V_s}{2 L_{lk} f_{sw}} D(1-D)$	P_o = output power f_{lk} = Leakage current V_p = Primary voltage V_s = Secondary voltage D = Duty cycle
Multi-element resonant converter	$N = D_{max} \dfrac{V_i}{V_o}$ = (Half-bridge) $N = D_{max} \dfrac{V_i}{2V_o}$ =(Full bridge)	D_{max} = Maximum Duty cycle V_o = Output voltage V_i = Input voltage
Push-pull converter model	$V_o = \dfrac{2DV_i}{n}$	D = Duty cycle V_o = Output voltage V_i = Input voltage n = Inverse of the transformer turns ratio

2.4 MAXIMUM POWER POINT TRACKING AND ITS CONTROL

In this section, the concept and need for MPPT along with modeling and characteristics of solar PV is discussed. Various types of MPPT schemes along with control strategies in DC–DC converters with examples are also elaborated. Few advanced control techniques for MPPT are also discussed.

2.4.1 Modeling and Characteristics of Solar PV

The basic elements of a solar PV system are PV cells. These cells are connected to form modules. It is further expanded in the form of arrays as per the power requirements as shown in Figure 2.26.

These PV cells exhibit nonlinear characteristics. The output of the PV cell varies with solar irradiation and with ambient temperature. The equivalent circuit model of the PV cell and module is shown in Figure 2.27a. The characteristic equation of the PV cell based on this model is given by equations (2.15–2.17) [15,16].

$$I = I_{ph} - I_{os} \{\exp [(q/AKT) (V + I\,R_s)] - 1\} - ((V + I^*R_s)/R_p) \tag{2.15}$$

$$I_{os} = I_{or} \exp [q\, E_{GO} / Bk\, ((1/T_r) - (1/T))]\, [T/T_r]3 \tag{2.16}$$

$$I_{ph} = S[\, I_{sc} + KI\, (T-25)]/100 \tag{2.17}$$

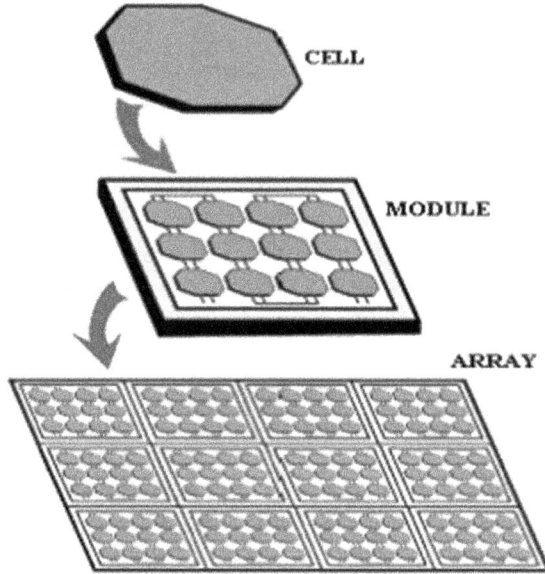

FIGURE 2.26 Solar PV configuration.

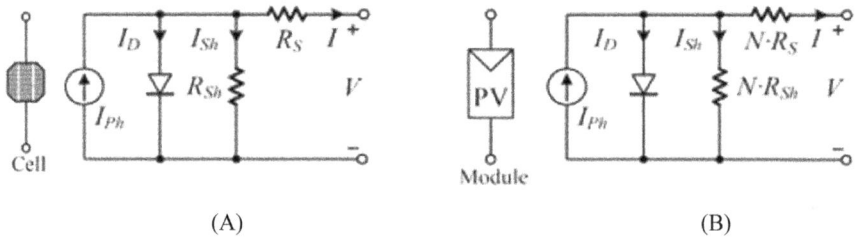

(A)   (B)

FIGURE 2.27 Equivalent circuit model of (a) solar PV cell and (b) solar PV module.

The current equation of the PV module where N identical cells are connected in series (Figure 27b), is given by equation 2.18.

$$I = I_{ph} - I_{os} \{\exp [(q/AKT)(V+I*N*R_s)] - 1\} - ((V+I*N*R_s)/(N*R_p)) \quad (2.18)$$

where I is the PV module output current, V is the PV cell output voltage, R_p is the parallel resistor, R_s is the series resistor, I_{os} is the PV module reversal saturation current, A and B are ideality factors, T is the temperature (°C), k is Boltzmann's constant, I_{ph} is the light-generated current, q is the electronic charge, K_I is the short-circuit current temperature coefficient at I_{SC}, S is the solar irradiation (W/m^2), I_{SC} is the short-circuit current at 25°C and 1000 W/m^2, E_{GO} is the bandgap energy for silicon, T_r is the reference temperature, and I_{or} is the saturation current at temperature T_r.

Figure 2.28 shows the variation of PV power with PV module voltage at different solar irradiation. The power increases with the increase in solar irradiation. Figure 2.29 shows the variation of PV module current with PV module voltage. Figure 2.30 is the variation of PV module power at different ambient temperatures. In all three figures, the variation is nonlinear. Also, the PV power decreases with the increase in ambient temperature.

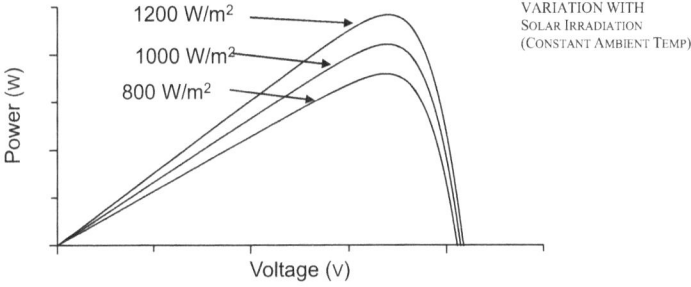

FIGURE 2.28   PV Characteristic curve of a module: power vs voltage.

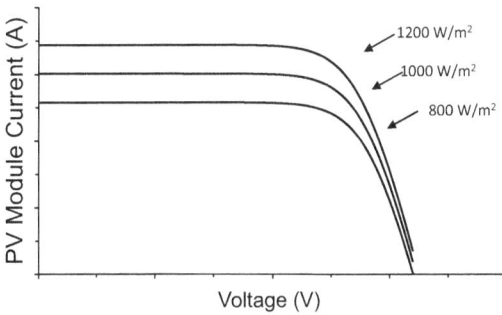

FIGURE 2.29   PV Characteristic curve of a module: current vs voltage.

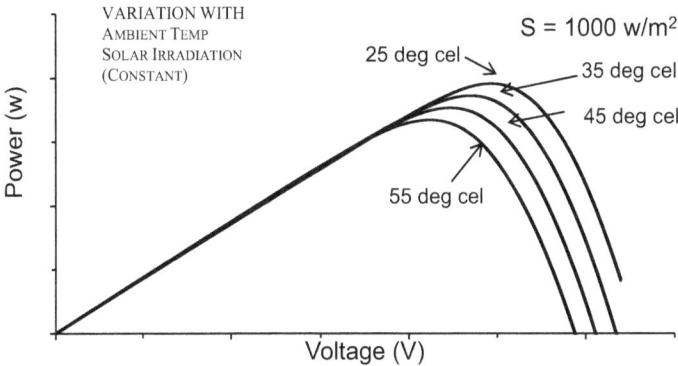

FIGURE 2.30   PV Characteristic curve of a module: power vs voltage.

2.4.2 MAXIMUM POWER POINT AND OTHER PARAMETERS OF PV CHARACTERISTIC

The solar PV I-V characteristic curve with parameters is shown in Figure 2.31. I_{sc} is the current of the PV module measured after shorting its terminals, i.e., it is known as short-circuit current. V_{oc} is the voltage measured at the output terminals of the PV module under open conditions. MPP is the maximum power point on the curve, i.e., it is the maximum power PV can deliver under an atmospheric condition. The voltage and current corresponding to MPP are denoted as V_{MPP} and I_{MPP} respectively. These parameters are provided by the manufacturer in its datasheet [17,18].

Figure 2.32 shows the variation of MPP of the solar PV curve under different solar irradiation. It increases with the increase in solar irradiation. Under any ambient condition, it is essential to harness maximum power from solar PV due to its slow conversion efficiency. A method known as MPPT is used to harness maximum power from solar PV. The concept of MPPT and its various types proposed by researchers are discussed in detail in the upcoming section.

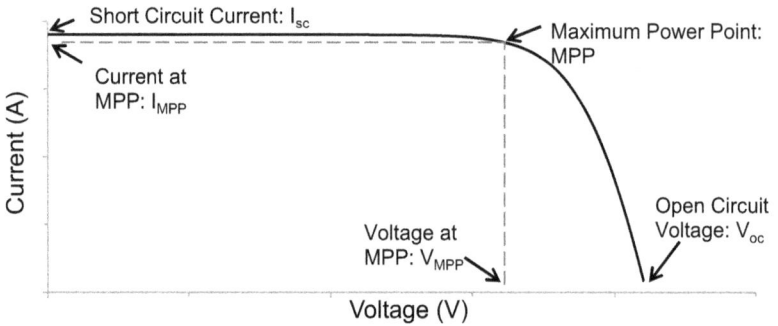

FIGURE 2.31   PV Characteristic curve parameters: current vs voltage.

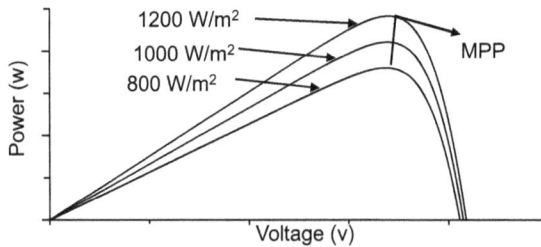

FIGURE 2.32   PV Characteristic curve: variation of MPP.

2.5   OVERVIEW OF VARIOUS MPPT TECHNIQUES

This section describes various MPPT techniques proposed for solar PV energy conversion systems along with their algorithm. These techniques include open-circuit voltage, short-circuit current, incremental conductance, perturb and observe, beta method, temperature method, hybrid method, artificial neural network, etc. Few of the techniques are evaluated for comparison [35].

The MPPT techniques are broadly classified into three categories, i.e.,

   i. Offline MPPT Methods
  ii. Online MPPT Methods
 iii. Other MPPT Methods.

The classification is shown in Figure 2.33.

A detailed description of these techniques has been discussed in the following sections.

2.5.1   Offline MPPT Techniques

Offline MPPT control techniques use technical data of PV panels to estimate MPP. These data include prior information like the I-V curve, P-V curve, and its variation with solar irradiation and other ambient conditions.

The following types of MPPT techniques come under this category.

2.5.1.1   Short-Circuit Current Technique

This method is the simplest MPPT method. There is an empirical relationship between short-circuit current of solar PV, i.e., I_{SC} and current at MPP, i.e., I_{MPP} given in equation 2.19:

$$I_{MPP} \approx K * I_{SC} \tag{2.19}$$

K is a constant which varies from 0.8 to 0.9. The short-circuit current (SCC) method gives accurate results under one ambient condition. The issue with this method is load is to be disconnected with solar PV to measure the short-circuit current. Also, the measurement of I_{SC} needs special arrangements, particularly for high currents which makes this method costly. Few researchers have used the switch used in power electronics converter to short the solar PV panel and measure the I_{SC} like using power metal–oxide–semiconductor field-effect transistor (MOSFET) used in boost DC–DC converter. Figure 2.34 is the flow chart of this technique.

2.5.2   Open-Circuit Voltage Technique

This method uses the empirical relationship between the voltage at MPP, i.e., the V_{MPP} and open-circuit voltage of solar PV V_{OC} given by equation 2.20.

$$V_{MPP} \approx K * V_{oc} \tag{2.20}$$

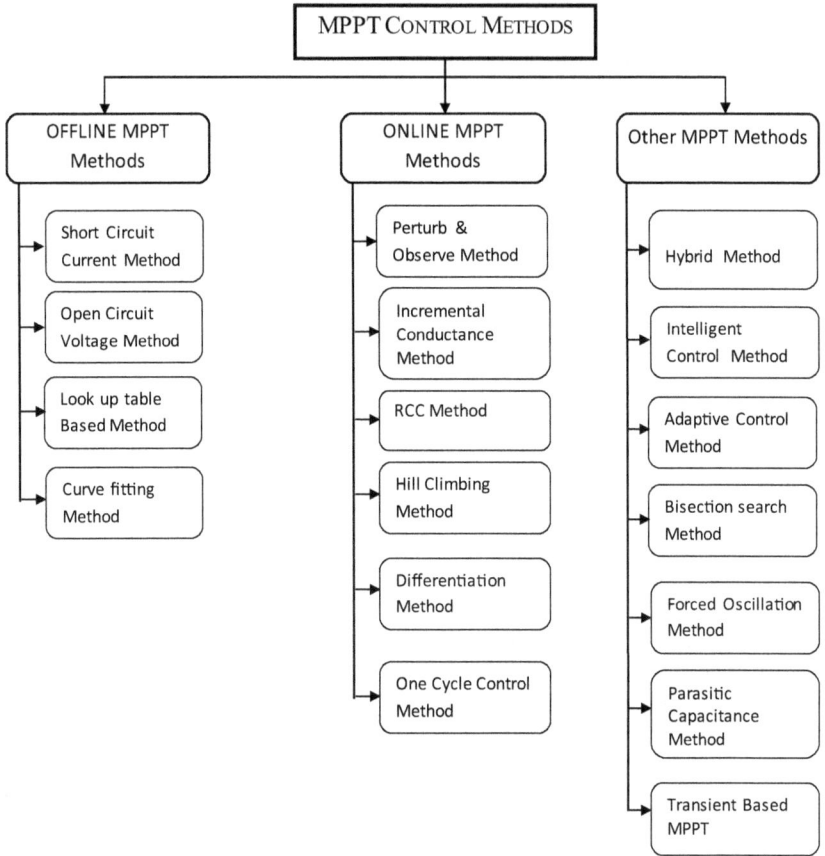

FIGURE 2.33   Classification of MPPT control methods.

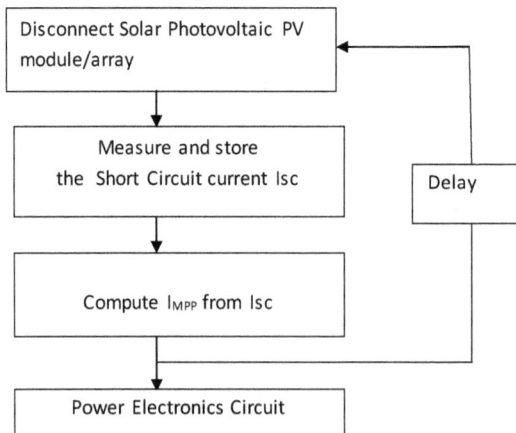

FIGURE 2.34   Flowchart diagram of short-circuit current method.

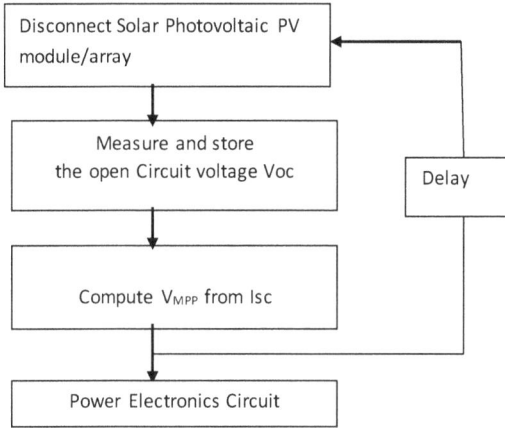

FIGURE 2.35   Flowchart diagram of open-circuit voltage method.

where K is a constant which depends on the solar cell characteristics. This relationship is based on the empirical results after the measurement of V_{oc} and V_{MPP}. It is reported that K varies from 0.73 to 0.8 for polycrystalline PV modules.

This method is an offline method. Every time V_{oc} is to be measured at the solar PV panel after disconnecting it from the load and then the signal is sent to the controller for computing V_{MPP}. This method suffers from the disadvantage that MPP may not be tracked accurately and every time the power to the load will be interrupted as V_{oc} can be measured by disconnecting the load from the solar PV panel, due to which power loss will increase. To overcome this issue pilot PV cells may be used to obtain V_{oc} with the care that the characteristic of the PV array is represented realistically. The other solution to this problem is to develop a realistic model by measurement of solar irradiation and ambient temperature to calculate V_{OC}. The flow chart of this MPPT technique is shown in Figure 2.35.

Both the open-circuit voltage and short-circuit current MPPT method have the advantage of fast-tracking and one loop control system. Both have the disadvantage that it is unable to track accurately under changing atmospheric conditions. It is a very suitable method for ambient conditions like space.

2.5.3   LOOK-UP TABLE-BASED TECHNIQUE

In this method, prior knowledge of solar PV panel data is required. The variation of V_{MPP}, I_{MPP} with ambient conditions is collected from the vendor's datasheet. These data are stored in the memory of the controller. The PV voltage and currents are sensed and compared with the stored data in memory. This method has the drawback that it requires a large amount of data storage memory and tracking speed is also reduced.

2.5.4   CURVE FITTING TECHNIQUE

In this offline technique method, the data of PV panel parameters and their variation along with an ambient temperature is required. In addition, the output characteristic equations are decided in advance. Based on these characteristic equations and mathematical model, voltage of MPP is calculated and the operating point on the characteristics of the panel is moved to track the MPP.

One such model is given by equation 2.21:

$$P_{pv} = A V_{pv}{}^3 + B\ V_{pv}{}^2 + C\ V_{pv} + D \tag{2.21}$$

The coefficients A, B, C, and D are calculated based on the sampling value, PV voltage, current, and power, respectively. After the calculation of these coefficients, the voltage at MPP can be calculated using equation 2.22:

$$V_{MPP} = (-B\ +/-\ \text{sqrt}\ (B^2 - 3AC))/3A \tag{2.22}$$

Simplicity is the advantage of this method. A large amount of data storage is required and tracking speed is reduced due to computations, which are the main disadvantages of this method. The values of the coefficients will change under changing ambient conditions.

2.6   ONLINE MPPT TECHNIQUES

In this category, the MPP is achieved by online tracking of PV power. The real-time data of current and voltage is used to track MPP. In these techniques, prior knowledge of PV panel data from the manufacturer is not required.

The following types of MPPT methods come in online MPPT techniques.

2.6.1   PERTURB AND OBSERVE (P&O) TECHNIQUE

The perturb and observe (P&O) method is another MPPT scheme. In the P&O method, the perturbation is applied either in the reference voltage or in the reference current signal of the solar PV. The flow chart of the P&O method is shown in Figure 2.36. In this chart, Y is shown as the reference signal. It could be either solar PV voltage or current. The main aim is to reach the MPP. To achieve it the system operating point is changed by applying a small perturbation (ΔY) in the solar PV reference signal. After each perturbation, the power output is measured. If the value of power measured is more than the previous value then the perturbation in the reference signal is continued in the same direction. At any point, if the new value of solar PV power is measured less than the previous one then perturbation is to apply in the opposite direction. This process is continued till MPP is reached. the P&O method uses the solar PV panel current as a reference signal. The issue with this method is that it becomes oscillatory around MPP. The MATLAB-based modeling is given for MPPT.

The other related issues of this method and solutions provided by other researchers are discussed in the upcoming section in detail. Summary of the P&O MPPT method is given in Table 2.3.

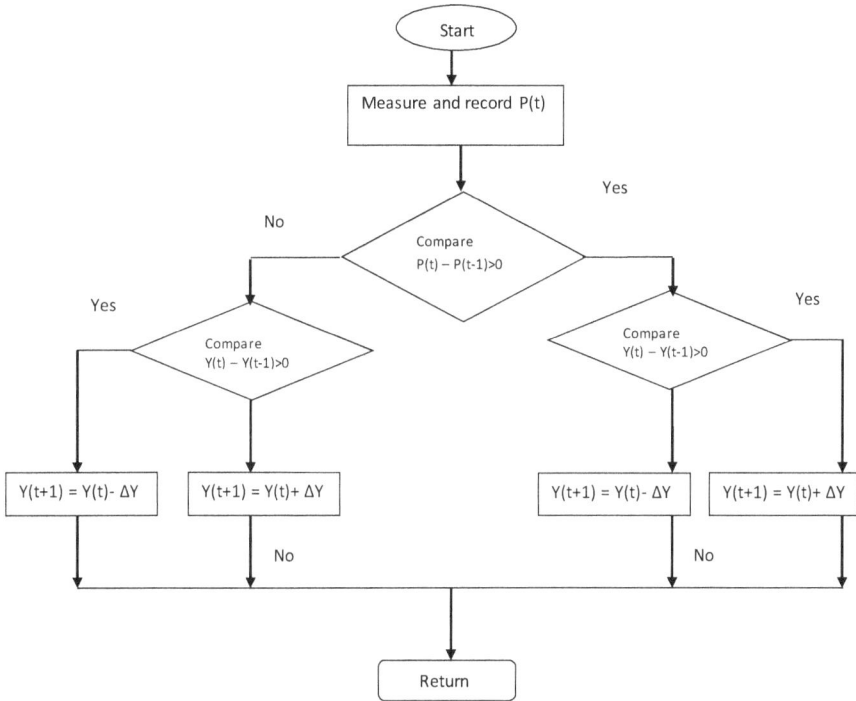

FIGURE 2.36 Flowchart diagram of the conventional P&O MPPT method.

TABLE 2.3
Summary of the Perturb and Observe MPPT Method

Perturbation	Change in Power	Next Perturbation
Positive	Positive	Positive
Positive	Negative	Negative
Negative	Positive	Negative
Negative	Negative	Negative

2.6.2 Incremental Conductance (IC) Technique

The incremental conductance (IC) method is a scheme of MPPT. The IC method is a widely accepted method. This MPPT scheme is based on the principle that the power slope of the PV characteristic curve at MPP is zero, i.e.,

$$dP/dV = 0 \tag{2.23}$$

$$dP/dV = d(V*I)/dV = I + V \, dI/dV \tag{2.24}$$

$$\Delta I/\Delta V = -I/V \tag{2.25}$$

TABLE 2.4

Summary of the IC MPPT Method

Incremental Conductance	Compare with Conductance at MPP	Conclusion	Action
$\Delta I/\Delta V$	$= I/V$	Operating point is the MPP	No Change
$\Delta I/\Delta V$	$>I/V$	Operating point is on the left of the MPP	Increase V_{ref}
$\Delta I/\Delta V$	$<I/V$	Operating point is the right of the MPP	Decrease V_{ref}

Equations 2.23–2.24 and Table 2.4 give the conditions of the IC method. At MPP, the instantaneous conductance and IC will have the same value. The IC method is used to track MPP by the sequence shown in a flowchart, i.e., Figure 2.37. As shown in the flowchart, V_{ref} is the voltage of solar PV at MPP. Once the MPP is tracked,

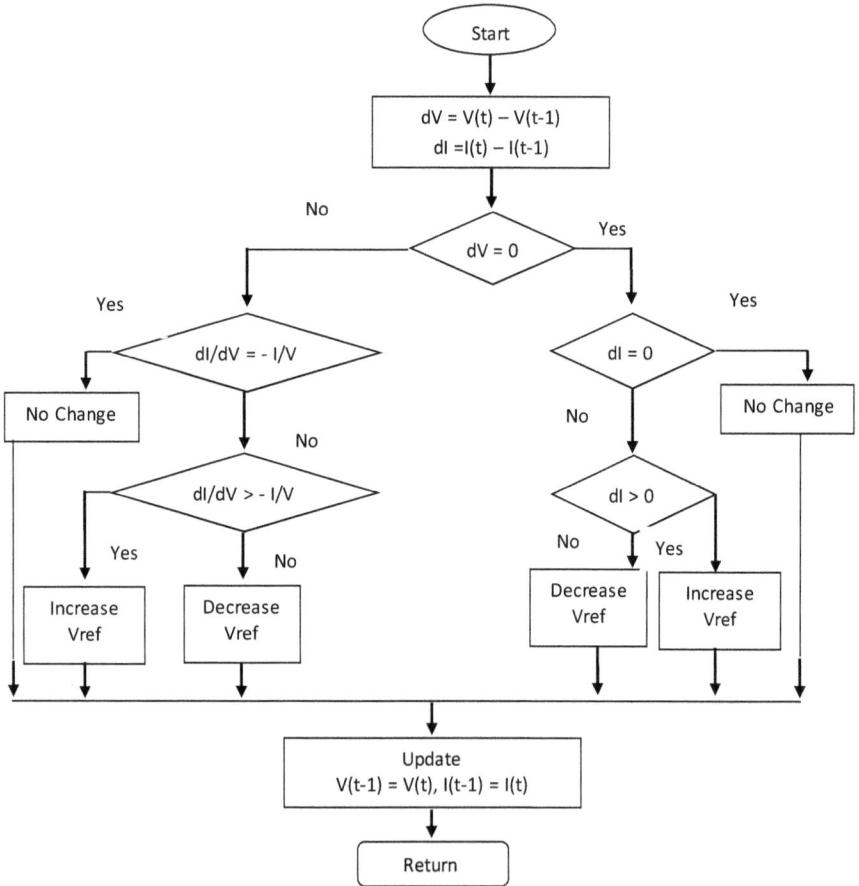

FIGURE 2.37 Flowchart diagram of the IC MPPT method.

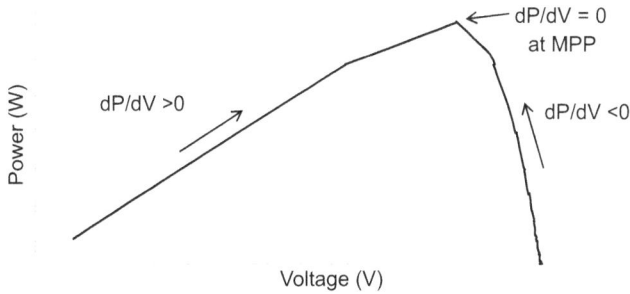

FIGURE 2.38   Solar PV curve for the IC MPPT technique.

the operating point is fixed at that point. If the working condition is changed, like load condition or environment condition, then the V_{ref} value is changed and again the instantaneous and IC are compared to get the new MPP. The change in V_{ref} value can be either side, i.e., increment or decrement.

The size of the change in V_{ref} decides the MPP tracking speed. The large change in V_{ref} will make this scheme faster; however, the large size change of V_{ref} may increase the oscillation around MPP. So a researcher has to make a compromise between speed and oscillations. This method has the advantage that it does not get affected by fast-changing atmospheric conditions. However, the challenge for the engineers is to reduce the complexity of the circuit.

Figure 2.38 shows the PV curve for the incremental conduction method. At MPP, the slope $dP/dV=0$. The duty cycle of the power electronics converter is changed to change the reference level for tracking MPP. The issue of fixed step size and low convergence time can be resolved using the variable step-size IC MPPT method. Villalva has proposed a small signal modeling of the IC method with a boost DC–DC converter.

IC MPPT is implemented with a pumping system and using an Alf and Vegard's Reduced (AVR) instruction set computer processor microcontroller. The results are elaborated for change in solar irradiation only. The IC MPPT is implemented using interleaved boost converter and shows improved results. A comparative simulation study is conducted between P&O and IC methods. A variable step-size method is introduced, in which the slope of the PV curve is multiplied by a constant. In all these methods, reliability issues are not considered deeply.

2.7   OTHER MPPT TECHNIQUES

The other MPPT techniques include either combination or modification of the offline and online methods. The modification is based on indirect calculations.

The following MPPT technique comes under this category.

2.7.1   Hybrid Technique

In the previous sections, offline and online MPPT methods are discussed with their advantages and disadvantages. The other category of MPPT techniques proposed by researchers are hybrid MPPT methods. In this method, both offline and online techniques are used in combination.

The hybrid method of the MPPT algorithm consists of two parts of its control signal. The first part of it is determined using any of the offline methods, which mainly depends on the atmospheric conditions of the solar PV panel and gives the fixed steady-state value. This part is required to give a fast response to environmental variations.

The second part of the control signal, which can be obtained by using any of the online methods, has the role to track MPP exactly. In contrast to the first part of the control signal, this part attempts to minimize the steady-state error and does not require a fast response to the environmental variations.

In this method, MPP is achieved in two steps. In the first step, the open-circuit voltage and short-circuit MPPT methods are used by multiplying open-circuit voltage V_{OC} with a factor k discussed in the previous section and short-circuit factor with a factor k. This step brings the operating point near MPP after which the second step is introduced by applying the P&O MPPT technique. The second step causes the fine-tuning of the operating point, i.e., reducing steady-state oscillations. A similar two-step hybrid MPPT method is recommended by using the IC method.

The flowchart diagram of this MPPT method is shown in Figure 2.39.

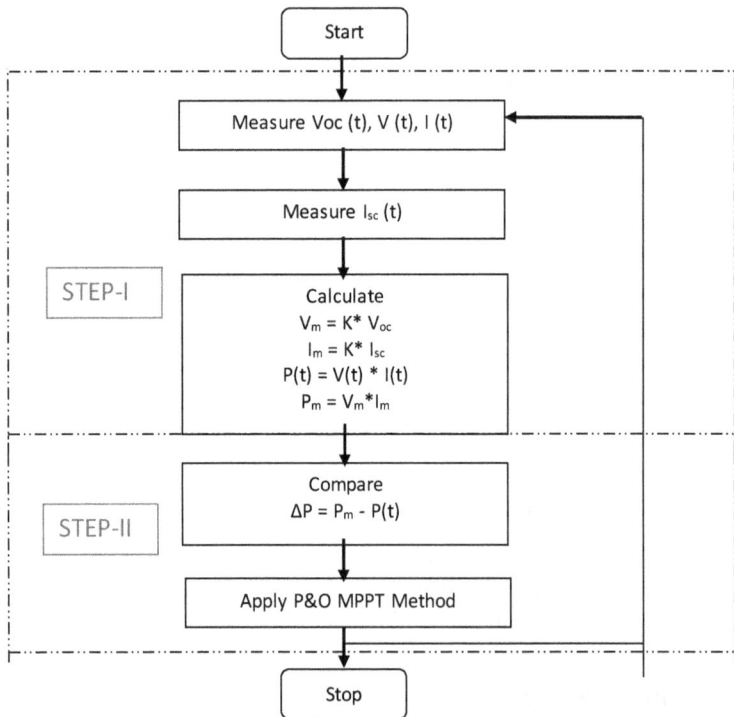

FIGURE 2.39   Flowchart diagram of the hybrid P&O MPPT technique.

This method gives faster results as compared to conventional P&O and IC MPPT methods. The drawback of this method is the interruption of load with the PV system to measure open-circuit voltage and short-circuit current. This interruption will lead to power interruption to the load, which is highly undesirable.

An ambient temperature should be measured to sense the change in atmospheric conditions. The aim of this solution is to avoid the interruption of load with PV for the measurement of open-circuit voltage and short-circuit current. The drawback of this solution is that an additional sensor is required, and the routine calibration of the thermal sensor may increase the cost.

The other solution is provided as an additional PV panel (along with the PV panel used to supply the power), whose V_{oc} and I_{sc} data can be used to sense the change in atmospheric conditions. It helps to avoid the interruption of load, but an additional PV panel and measurement setup will increase the cost and complexity of the system. Murtaza in reference [36] uses a combination of P&O and open-circuit voltage, but its variation with load and reliability are not considered.

2.7.2 INTELLIGENT CONTROL TECHNIQUE

This method can be further divided into three categories:

   i. Fuzzy logic-based technique
  ii. Neural network-based technique
 iii. Optimization-based technique

The intelligent control-based method has the advantage of working with variable inputs, without the need for a mathematical model and has the ability to handle nonlinearity.

2.7.2.1 Fuzzy Logic-Based Technique

The fuzzy logic controller varies the duty cycle based on the error in the input signal. The instant at which the error becomes zero means that the panel output voltage is equal to the MPP voltage. The error signal is obtained by comparing the instantaneous panel voltage with a reference voltage. The reference voltage is MPP voltage at solar irradiation. The input variable could be taken as solar PV power.

The diagram of the fuzzy logic-based MPPT method is shown in Figure 2.40. The fuzzy logic method works in three steps:

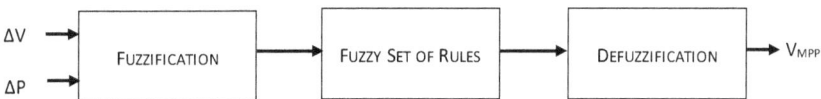

FIGURE 2.40 Diagram of the fuzzy logic-based MPPT technique.

Step I: Fuzzification

Fuzzification is a process in which the error signal or change in error signal is converted into linguistic variables. Then, the membership function is chosen between [−1 1]. In this method, two variables, i.e., power and voltage, are used to express the fuzzy set of rules.

Step II: Fuzzy Set of Rules

A set of fuzzy rules are defined based on the system input and output behavior in the form of IF – THEN. A rule-based table is obtained with several possible combinations, to satisfy different conditions.

Step III: Defuzzification

In this process, the linguistic variables are converted to numerical values as the output based on the membership function and rules defined.

This method is fast, but the drawback is prior knowledge of MPP voltage under different conditions is required. Also, the circuit complexity is high to implement this method.

2.7.2.2   Neural Network-Based Technique

The availability of new embedded system tools like microcontrollers, DSP, etc. has made the artificial neural network (ANN) method popular for MPPT among the researchers in the last few years. The structure of the ANN is shown in Figure 2.41. The ANN commonly has three layers: input, hidden, and output layers. The number of nodes in each layer varies and is user-dependent. For solar PV, the input variables can be open-circuit voltage, short-circuit current, solar irradiation, ambient temperature, or a combination of these. The output could be one or more reference signal(s) like the duty cycle which is used to drive the power converter to operate at or close to the MPP. The availability of new embedded system tools like microcontrollers, DSP, etc. has

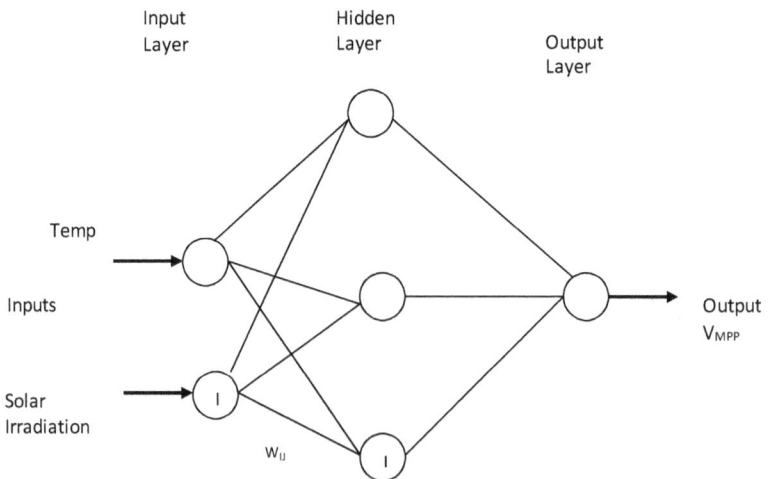

FIGURE 2.41   Diagram of the ANN-based MPPT technique.

made the ANN method popular for MPPT among researchers in the past few years. The structure of the ANN is shown in Figure 2.41. The ANN commonly has three layers: input, hidden, and output layers. The number of nodes in each layer varies and is user-dependent. For solar PV, the input variables can be open-circuit voltage, short-circuit current, solar irradiation, ambient temperature, or a combination of these.

The output could be one or more reference signal(s) like the duty cycle which is used to drive the power converter to operate at or close to the MPP. The closeness of the operating point with MPP depends on the use of the hidden layer algorithm and on the effective training of the neural network. The node links are weighted as shown in Figure 2.41; the link between node I and node J is labeled with weight W_{IJ}. These weights have to be carefully determined through the training process. Therefore, the PV array is tested for months and years and the pattern between the input and the output of the neural network is recorded. Hiyama was the first to propose the use of the ANN for MPPT. In his method, open-circuit voltage is taken as the only input signal which can be compared with the instantaneous voltage in order to generate the control signal required to derive the solar panel for MPP through a PID controller.

The advantage of the ANN method-based MPPT is that the trained neural network can provide a sufficiently accurate MPPT without requiring detailed knowledge of the solar PV parameters. However, it is to be kept in mind that different PV arrays have different characteristics; a neural network has to be specifically trained for the PV array with which it will be used. Furthermore, the characteristics of solar PV arrays also change with time, implying that the neural network has to be trained periodically for getting accurate MPPT.

2.7.2.3   Optimization-Based Technique

Since the MPP is a maxima point of the PV characteristic curve, which is solved using various optimization techniques like particle swarm intelligence (PSI), etc. These methods are used where multiple maxima are found. Other optimization techniques are also proposed by researchers for the MPPT method.

In reference [37], Femia has proposed an optimal control of centralized MPPT defining inequalities for the duty cycle and it is compared with distributed MPPT. Consideration of load variation is missing in this work. Wang in reference [38] has proposed a prediction method in which the resistance effect is considered on the I-V curve of PV. The calculation is less and gives faster results, but the reliability is an issue and load variations are not considered. In reference [39], the PSO algorithm is used for common MPPT with multiple PV arrays.

2.7.3   Adaptive Control Technique

The adaptive control method is used to get a variable perturbation size under changing atmospheric conditions. The block diagram of the adaptive MPPT method is shown in Figure 2.42. The adaptive method automatically tunes the perturbation step size large when the change in power becomes large due to the change in solar irradiation.

An adaptive controller reduces the step size when the change in power is low with respect to the pre-set value.

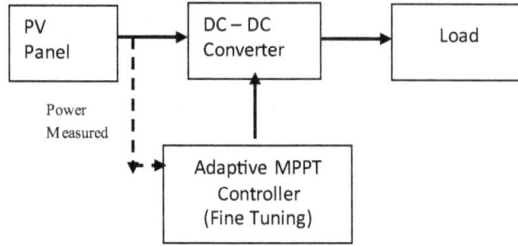

FIGURE 2.42   Diagram of the adaptive MPPT technique.

In this type, an adaptive MPPT method for P&O is proposed. Another adaptive and predictive P&O MPP method is proposed. The drawback of this method is that the emphasis is more on perturbation size and less on V_{MPP}.

2.8   CONTROL OF THE DC–DC CONVERTER FOR MPPT

Control of the DC–DC converter for MPPT operation depends on the type of solar PV plants, in which the DC–DC converter is used. The constraints and limitations will vary with the type of solar PV plant used. In this section, it is discussed in detail.

2.8.1   CONTROL OF THE DC–DC CONVERTER FOR MPPT
IN STANDALONE SOLAR PV-BASED PLANTS

The MPP is the track of point on the solar panel PV curve where maximum power can be extracted as discussed in Section 2.4. This supplied power level depends on solar irradiation, ambient temperature, and load. The duty cycle of the signal given to the gate of the DC–DC converter switch will vary to reach the MPP. At the same time, the change in power extracted from the solar panel is dependent on two following factors:

1. Due to changes in ambient conditions, i.e., solar irradiation, etc.
2. Due to changes in load conditions.

The controller of the DC–DC converter should be capable of distinguishing the cause of PV power change and executing accordingly. The summary of this change is given in Table 2.5:

Based on the above table, the MPP control region on the PV curve varies with the type of the DC–DC converter.

TABLE 2.5:
Cause and Effect of Change

Cause	Change	Power	Voltage
Solar Irradiation	Increase	Increase	Increase
	Decrease	Decrease	Decrease
Load Resistance	Increase	Decrease	Increase
	Decrease	Increase	Decrease

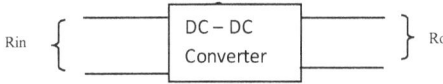

FIGURE 2.43   Block diagram of a DC–DC converter.

2.8.2   DC–DC CONVERTER TOPOLOGY AND MPPT WORKING ZONE

The MPPT operating zone for solar PV is dependent on DC–DC converter topology and restricts the value of resistive load for which MPPT become effective. The MPPT scheme is implemented in the control of the DC–DC converter; i.e., it varies the duty cycle. The basic principle of adjusting the duty cycle is to match load impedance with input impedance seen by the DC–DC converter, i.e., the impedance of solar PV as shown in Figure 2.43. R_{in} (the input impedance seen by the converter) and R_o (the output impedance connected to the converter) are related to the characteristic equation.

This mathematical equation varies with DC–DC converter topologies, which are summarized in Table 2.6. Table 2.7 shows the operating and non-operating zone of MPPT. The MPPT works in the entire range of characteristic curves in the case of buck-Boost and the SEPIC DC–DC converter. Since these converters are complex, they exhibit more cost in comparison to buck or boost converters.

2.8.3   OPTIMIZATION OF THE DUTY CYCLE FOR MPPT

The most commonly used MPPT technique is the hybrid MPPT technique. In the hybrid MPPT technique, two types of MPPT techniques are clubbed to achieve MPP. In this section, a method to calculate the optimized duty for reaching MPP is described.

The theme of the hybrid MPPT is shown in Figure 2.44.

The hybrid technique design is done in three phases [40]. In first phase, the fractional open-circuit voltage (FOCV) is implemented.

In the PV system, any type of DC–DC converter can be used. The buck DC–DC converter is used in this design; however, the same technique is valid for all other types of DC–DC converters.

The PV array reaches its MPP when the output load (R_{out}) becomes equal to the internal impedance $R_{imp,pv}$ of the PV array. With different atmospheric conditions, the $R_{imp,\,pv}$ of the PV array changes all the time. However, R_{out} does not change according to ambient conditions or on its own. The relation between the input voltage (V_{in}) and output voltage (V_{out}) of a buck converter is:

$V_{out} = D\ V_{in}$, where D is the duty cycle. With 100% efficiency, $P_{in} = P_{out}$; therefore

$$V_{in}\ I_{in} = V_{out}\ I_{out}$$
$$(V_{in})2/R_{in} = (V_{out})2/R_{out}$$

which leads to

$$R_{in} = (1/D^2)\ R_{out}$$

R_{out} and R_{in} are the two parameters of the converter. R_{out} remains the same but load seen by the PV module, i.e., R_{in} can be varied with the help of D.

Actually, by changing D, R_{in} is varied, and in response, the PV module changes its operating voltage (V_{pv}). It is the responsibility of the MPPT algorithm to set D at such an optimized value (D_{mpp}) that R_{in} matches the internal impedance of the array

TABLE 2.6

Characteristic Equations of Commonly Used DC –DC Converters

Characteristic Equation of Buck Converter

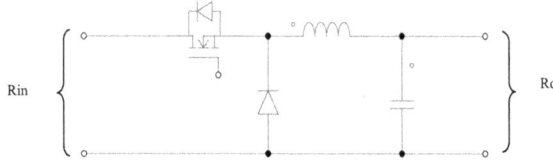

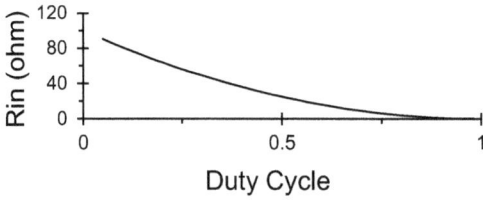

Rin

Ro

Input output
relationship
$R_{in} = R_o/D^2$
D – Duty cycle

Rin (ohm) vs **Duty Cycle**

Rin vs Duty Cycle

Characteristic Equation of Boost Converter

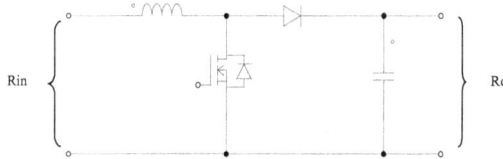

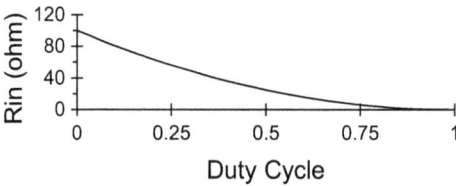

Rin

Ro

Input output
relationship
$R_{in} = R_o(1-D)^2$
D – Duty cycle

Rin (ohm) vs **Duty Cycle**

Rin vs Duty Cycle

Characteristic Equation of Buck-Boost Converter

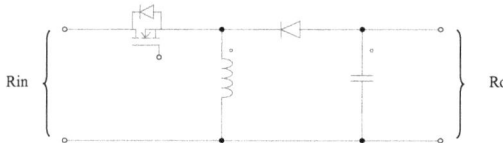

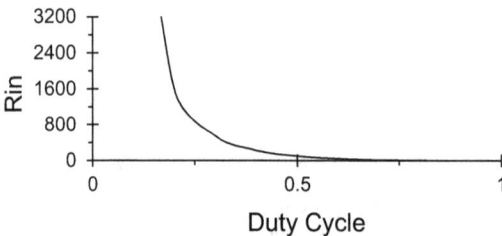

Rin

Ro

Input output
relationship
$R_{in} = (1-D)^2 * R_o/D^2$
D – Duty cycle

Rin vs **Duty Cycle**

Rin vs Duty Cycle

(Continued)

TABLE 2.6 (*Continued*)

Characteristic Equations of Commonly Used DC –DC Converters

Characteristic Equation of SEPIC Converter

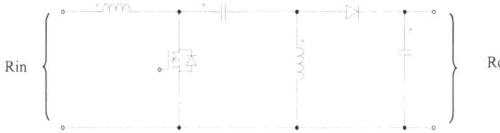

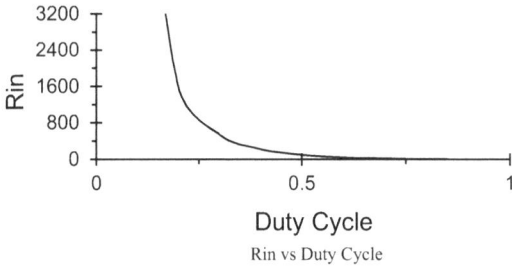

Input output
relationship
$R_{in} = (1 - D)^2 * R_o/D^2$
D – Duty cycle

Rin vs Duty Cycle

TABLE 2.7

MPPT Working Zone with DC–DC Converter Topology

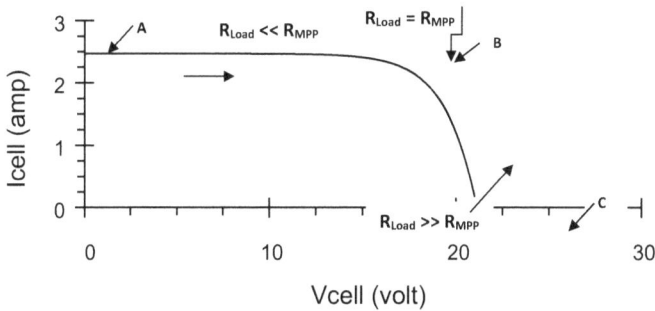

V – I Characteristic Curve of Solar PV

V–I Characteristic Curve of Solar PV

DC–DC Converter Type	MPP Zone	
	Working Zone	No Working Zone
Buck	A-B	B-C
Boost	B-C	A-B
Buck-Boost	A-B, B-C	None
SEPIC	A-b, B-C	None

(R_{imp}, pv). At this stage, the module starts operating at the MPP voltage (V_{mpp}) which delivers the maximum power.

At non-MPP. $R_{in} = R_{out}/D^2$ $R_{in} \neq R_{imp}$, pv; $V_{pv} \neq V_{mpp}$
At MPP. $R_{in=} R_{out}/D^2$ mpp $R_{in} = R_{imp}$, pv; $V_{pv} = V_{mpp}$

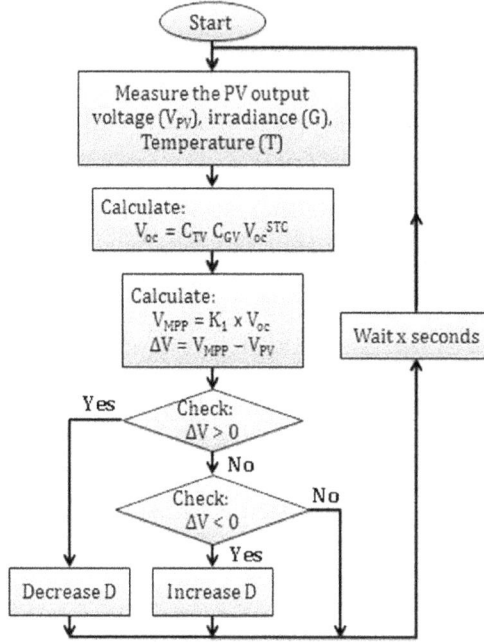

FIGURE 2.44   Fractional open circuit voltage-based MPPT.

Expression for Optimized Duty Cycle (D):
In the first step, solar PV voltage

$$V_{pv} = k \, V_{oc},$$

where k is the constant, and V_{oc} is the open-circuit voltage of the PV panel.

The expression for optimized duty cycle D formulas is derived for the buck converter. However, it can be derived for any DC–DC converter using the same procedure. When the PV array does not operate at the MPP, the load seen by the PV module R_{in} can be written as follows:

$$R_{in} = R_{out}/D^2 \quad R_{out} = D^2 \, R_{in}$$

Assume that the output load (R_{out}) is constant. When the PV array operates at the MPP with the optimized duty cycle (D_{mpp}), its internal impedance becomes equal to R_{in}, i.e., $R_{in} = R_{imp, pv}$. Hence, $R_{imp, pv}$ can be written as follows:

$$R_{imp, pv} = R_{out}/D^2_{mpp}$$
$$R_{out} = D^2_{mpp} \, R_{imp, pv} \qquad\qquad (2.26)$$

By combining

$$D^2_{mpp} \, R_{imp, pv} = D^2 R_{in}$$

$$D_{mpp} = \sqrt{\dfrac{D^2 \, R_{in}}{R_{imp, pv}}}$$

$R_{imp, pv}$ is the scenario when the PV array is operating at the MPP, so we can find it through the relation

$$R_{imp, pv} = V_{mpp}/I_{mpp}$$

$$R_{imp, pv} = kV_{oc}/I_{mpp}$$

R_{in} is the case when the PV array is not operating at the MPP, so it can be found out as

$$R_{in} = V_{pv}/I_{pv}$$

Putting these values together, we can have:

$$D_{mpp} = \sqrt{\frac{D^2 V_{pv} \, I_{mpp}}{KV_{oc} \, I_{pv}}}$$

I_{sc} is close to I_{mpp}, we can assume that $I_{pv} = I_{sc} \approx I_{mpp}$. Using the above equations, D-optimization formula for region-1 can be written as:

$$D_{mpp} = \sqrt{\frac{D^2 V_{pv}}{kV_{oc}}} \tag{2.27}$$

On the other hand, if we are operating in a region greater than kV_{oc} i.e. region-2, then a simple slope-line formula is used to calculate I_{mpp} as shown in Fig. 2.45. First from the known values of V_{pv} and I_{pv} of operating point, we can write slope relation between V_{oc} point and operating point as:

$$Slope = \frac{0 - I_{pv}}{V_{oc} - V_{pv}} \tag{2.28}$$

After getting slope, we can write again two-point slope relation between V_{oc} point and MPP as:

$$I_{mpp} = \frac{I_{pv}(V_{oc} - kV_{oc})}{V_{oc} - V_{pv}} \tag{2.29}$$

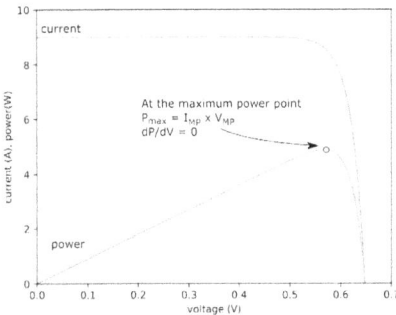

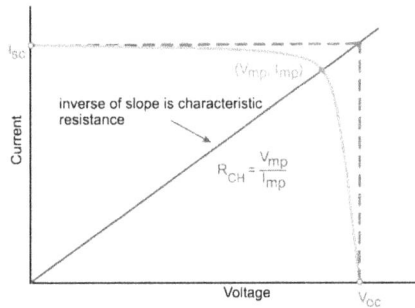

(a) The maximum power voltage occurs when the differential of the power produced by the cell is zero.

(b) characteristic resistance

FIGURE 2.45 Slope-line approach.

Using I_{mpp} from Eqs. (2.29) in (18), D-optimization formula for region-2 can be written as:

$$D_{mpp} = \sqrt{\frac{D^2 V_{pv}(V_{oc} - kV_{oc})}{kV_{oc}(V_{oc} - V_{pv})}} \qquad (2.30)$$

2.8.3.1 Other Factors to Be Considered in MPPT Design

1. Gap Between Two Perturbations of Duty Cycle

The perturbation will be applied in terms of the duty cycle of the Buck converter to achieve MPP. If the two perturbations are applied without giving the appropriate delay, then the MPPT may give false results, i.e., the perturbation should not be applied before the output settles down from the previous perturbation. The following factors are to be considered for deciding the gap between two perturbations:

    I. The transient response of buck DC–DC converter.

    II. Processing time taken by the microcontroller.

    III. Processing time taken by the voltage and current sensor.

2.9 QUESTIONS

1. What is a DC–DC converter?
2. What is the role of a DC–DC converter in different applications?
3. What is the role of a DC–DC converter in solar PV applications?
4. What is solar PV?
5. How is solar PV modeled?
6. What are the characteristics of solar PV?
7. What are the different parameters mentioned on solar PV characteristics?
8. How is the operation of solar PV linked to ambient conditions?
9. What is MPPT?
10. How does it work?
11. What are the different types of MPPT used?
12. What is P&O MPPT?
13. What is the IC MPPT technique?
14. What are offline MPPT techniques?
15. What are online MPPT techniques?
16. What are the characteristics of impedance?
17. What is the limitation of MPPT?
18. What is the role of the duty cycle in deciding MPPs?
19. What other factors are to be considered before deciding on MPPT techniques?
20. Why is MPPT important?

REFERENCES

[1] A. Haque, F. Blaabjerg, *Reliability of Power Electronics Converters*, IET UK, 2021.
[2] M. H. Rasheed, *Power Electronics-Devices and Circuit Applications*, 4th Edition, Pearson India, 2019.

[3] L. Gustavsson, P. Borjesson, B. Johansson, P. Svenningsson, "Reducing CO_2 emissions by substituting biomass for fossil fuels," *Int. J. Energy.*, vol. 20, pp. 1097–1113, 1995.

[4] M. Munster, H. Lund, "Use of waste for heat, electricity, and transport—challenges when performing energy system analysis," *Int. J. Energy.*, vol. 34, pp. 636–644, 2009.

[5] M. Beccali, P. Finocchiaro, B. Nocke, "Energy and economic assessment of desiccant cooling systems coupled with single glazed air and hybrid PV/ thermal solar collectors for applications in hot and humid climate," *Int. J. Sol. Energy.*, vol. 83, pp. 1828–1846, 2009.

[6] S. Mekhilef, R. Saidur, A. Safari, 2011. "A review on solar energy use in industries," *Renewable Sustainable Energy Rev.* Elsevier, vol. 15, pp. 1777–1790, 2011.

[7] K. Sakaki, K. Yamada, "CO_2 mitigation by new energy systems," *Energy Convers. Manag.* Elsevier, vol. 38, pp. 655–660, 1997.

[8] K. Yamada, H. Komiyama, K. Kato, A. Inaba, "Evaluation of photovoltaic energy systems in terms of economics, energy, and CO_2 emissions," *Int. J. Energy Convers. Manag.* Elsevier, vol. 36, pp. 819–822, 1995.

[9] P. Nema, R. K. Nema, S. Rangnekar, "Minimization of green house gases emission by using hybrid energy system for telephony base station site application," *Int. J. Renewable Sustainable Energy Rev.* Elsevier, vol. 14, pp. 1635–1639, 2010.

[10] P. K. Katti, M. K. Khedkar, "Alternative energy facilities based on site matching and generation unit sizing for remote area power supply," *Int. J. Renewable Energy.* Elsevier, vol. 32, pp. 1346–1366, 2007.

[11] G. Spagnuolo, G. Petrone, S. V. Araujo, "Renewable energy operation and conversion schemes," *IEEE Ind. Electron. Mag.*, vol. 4, pp. 38–51, 2010.

[12] O. Stalter, D. Kranzer, S. Rogalla, B. Burger, "Advanced Solar Power Electronics," 22nd *International Symposium on Power Semiconductor Devices & IC's*- IEEE, pp. 3–10, 2010.

[13] M. Ameli, S. Moslehpour, M. Shamlo, "Economical load distribution in power networks that include hybrid solar power plants," *Int. J. Electr. Power Syst. Res.* Elsevier, vol. 78, no. 7, pp. 1147–1152, 2008.

[14] G. Beghin, V. Nguyen, T. Phonoc, "Power conditioning unit for photovoltaic power systems," In *Proceedings of the 3rd European Communities Photo-Voltaic Solar Energy Conference*, pp. 1041–1045, 2010.

[15] R. Hernanz, C. Martin, Z. Belver, L Lesaka, G Zulueta, G. P. Pérez, "Modelling of photovoltaic module," In *Proceedings of the International Conference on Renewable Energies and Power Quality (ICREPQ'10).* Granada, Spain, pp. 23–25, March, 2010

[16] I. Houssamo, F. Locment, M. Sechilariu, "Maximum power tracking for photovoltaic power system: development and experimental comparison of two algorithms," *Int. J. Renewable Energy.* Elsevier, vol. 35, pp. 2381–2387, 2010.

[17] M. G. Villalva, J. R. Gazoli, E.R. Filho, "Comprehensive approach to modelling and simulation of photovoltaic arrays," *IEEE Trans. Power Electron.*, vol. 5, pp. 1198–2008, 2009.

[18] S. Solanki Chetan, *Solar Photovoltaics: Fundamentals, Technologies and Applications*, New Delhi: PHI, 2012.

[19] N. Wang, C. Hun, M. Y. Wu, G. Shi, S. Heng, "Study on character istics of photovoltaic cells based on MATLAB simulation," In *Proceedings of IEEE Power and Energy Engineering Conference* Asia-Pacific, pp. 1–4, 2011.

[20] D. Meneses, F. Blaabjerg, O. Garcia, J. A. Cobos, "Review and comparison of step up transformer less topologies for photovoltaic AC module application," *IEEE Trans. Power Electron.*, vol. 28, no. 6, pp. 2649–2663, 2013.

[21] J. D. Bollinger, K. Corzine, B. Chowdhury, M. Ferdowsi, *"Applications of Solar Energy to Power Standalone Area and Street Lighting"* [Master of Science in Electrical Engineering]. University of Missouri-Rolla; 2007.

[22] Ritchie, Ralph W. *Using Sunlight for Your Own Solar Electricity: Build Your Own System, Become Independent of the Grid, Domestic Photo Voltaics.* Springfield: Ritchie Unlimited Publications, 1999.

[23] S. Steven, S. William, *The Solar Electric House*, Sustainability Press, pp. 8–16, 1993.

[24] B. Shone, G. Stapleton, M. Russell, Wilmot Nigel, *Electricity from the Sun: Solar PV Systems*, 3rd Edition, Clean Energy Council, Australia, pp. 5–14, 2008.

[25] M. Davis, B. Dougherty, A. Fanney, "Short-term characterization of building integrated photovoltaic panels," *Int. J. Sol. Energy Eng.-ASME*, vol. 125, pp. 13–20, 2003.

[26] "IEEE Guide for Array and Battery Sizing in Stand-Alone Photovoltaic (PV) Systems," in IEEE Std 1562-2007, pp. 1–32, 2008. doi: 10.1109/IEEESTD.2008.4518937.

[27] Architectural Energy Corporation. Maintenance and operation of stand-alone photovoltaic systems. Photovoltaic Design Assistance Center, Sandia National Laboratories, 1991.

[28] F. I. Bakhsh, D. K. Khatod, "Application of variable frequency transformer (VFT) for grid inter connection of PMSG based wind energy generation system," *Int. J. Sustainable Energy Technol. Assess.* Elsevier, vol.8, pp. 172–80, 2008.

[29] Y. Bae, T. K. Vu, R. T. Kim, "Implemental control strategy for grid stabilization of grid-connected PV system based on German grid code in symmetrical low- to-medium voltage network," *IEEE Trans. Energy Convers.*, vol. 28, no.3, pp. 619–631, 2013.

[30] J. M. Chang, W. N. Chang, S. J. Chiang, "Single-Phase Grid-Connected PV. System using three-arm rectifier-inverter," *IEEE Trans. Aerosp. Electron. Syst.*, vol. 42, no. 1, pp. 211–219, 2006.

[31] L. C. Khanh, J. J. Seo, Y. S. Kim, D. J. Won, "Power-management strategies for a grid-connected PV–FC hybrid system," *IEEE Trans. Power Delivery*, vol. 25, no. 3, pp. 1874–1882, 2010.

[32] G. V. Quesada, F. G. Gispert, R. P. López, M. R. Lumbreras, A. C. Roca, "Electrical PV array reconfiguration strategy for energy extraction improvement in grid-connected PV systems," *IEEE Trans. Ind. Electron.*, vol. 56, no. 11, pp. 4319–4331, 2009.

[33] Y. Riffonneau, S. Bacha, F. Barruel, S. Ploix, "Optimal power flow management for grid connected PV systems with batteries," *IEEE Trans. Sustainable Energy* vol. 2, no. 3, pp. 309–320, 2011.

[34] S. Krithiga, G. A. Gounden, "Power electronic configuration for the operation of PV system in combined grid-connected and stand-alone modes," *IET Trans. Power Electron.*, vol. 7, no. 3, pp. 640–647, 2014. http://dx.doi.org/10.1049/iet-pel.2013.0107.

[35] M. Kane, D. Larrain, D. Favrat, Y. Allani, "Small hybrid solar power system," *Int. J. Energy*, Elsevier, vol. 28, pp. 1427–1443, 2003.

[36] A. F. Murtaza, H. A. Sher, M. Chiaberge, D. Boero, M. D. Giuseppe and K. E. Addoweesh, "A novel hybrid MPPT technique for solar PV applications using perturb & observe and Fractional Open Circuit Voltage techniques," Proceedings of 15th International Conference MECHATRONIKA, 2012, pp. 1-8.

[37] Femia, N., G. Petrone, G. Spagnuolo, and M. Vitelli. "Optimal Control of Photovoltaic Arrays." Mathematics and Computers in Simulation 91 (May 2013): 1–15. https://doi.org/10.1016/j.matcom.2012.05.002.

[38] Wang, Meng, Jinqing Peng, Yimo Luo, Zhicheng Shen, and Hongxing Yang. "Comparison of Different Simplistic Prediction Models for Forecasting PV Power Output: Assessment with Experimental Measurements." Energy 224 (June 2021): 120162. https://doi.org/10.1016/j.energy.2021.120162.

[39] M. Miyatake, M. Veerachary, F. Toriumi, N. Fujii and H. Ko, "Maximum Power Point Tracking of Multiple Photovoltaic Arrays: A PSO Approach," in IEEE Transactions on Aerospace and Electronic Systems, vol. 47, no. 1, pp. 367-380, January 2011, doi: 10.1109/TAES.2011.5705681.

[40] D. Baimel, S. Tapuchi, Y. Levron, J. Belikov, "Improved fractional open circuit voltage MPPT methods for PV systems," *Electronics*, vol. 8, p. 321, 2019. https://doi.org/10.3390/electronics8030321

3 Power Electronics Converter Designing. Part 2
AC Side

3.1 INTRODUCTION

With the increase in demand for power and the constant depiction of non-renewable resources, a solution for sustainable power generation is required. Power grid integration with a photovoltaic (PV) system is being implemented at a wide scale due to advances in technology, enabling clean and environment-friendly sources. For PV integration with the grid or even to operate a PV in stand-alone mode, an inverter setup with a DC–DC converter and controller is required. The DC–DC converter assists the PV panel in providing a stable DC input to the inverters. The inverter then converts the DC obtained from solar into AC. Inverter technology has come a long way in recent years. Previously, a separate transformer was used to provide galvanic isolation between the PV system and the grid. Transformers do provide physical isolation but also cause multiple losses in the system, causing a decline in the overall efficiency. The presence of the transformer also makes the inverter bulky. Because of the problems mentioned above, transformerless inverter topology emerges as a viable solution. In a transformerless inverter, the grid and PV system are interconnected with a filter in the middle to eliminate harmonics which was generated from the inverter. As the grid and the PV system are interconnected, an issue of leakage current arises because of the current flowing from the grid end to the PV system. To eliminate the issue of the leakage current, a new topology was designed. The new topologies consisted of extra power electronics switches. The switches provide a freewheeling mode to the inverter during positive-to-negative cycle switching and vice versa. The presence of freewheeling mode reduces the flow of leakage current toward the DC side of the circuit. There have been many different transformerless inverter topologies that have been developed over the years. In this chapter, a brief introduction regarding different transformerless inverter topologies is presented. Modulation techniques to switch are also explained in this chapter, which is followed by problem identification and newly proposed topology. Some basic simulation is presented in the chapter comparing the proposed topology with pre-existing topologies.

DOI: 10.1201/9781003257189-3

3.2 DISTRIBUTED GENERATION CONCEPT AND TECHNOLOGY

In this section, definitions of distributed generation (DG) as well as storage systems are given. Types of DG technology are presented.

3.2.1 DISTRIBUTED GENERATION CONCEPT

DG is a concept that existed years before. In the olden days, energy was provided by steam, hydraulics, direct heating, and cooling, and light and energy was produced near the device. These happened until electricity was introduced as an alternative for commercial purposes. From the point of the traditional view of planning and operation of distribution networks, the energy only flows from the transmission network, connecting generation sources across the state and country, through the distribution network, and to the customer, as shown in Figure 3.1.

The increasing adoption of new-generation technologies by customers presents technical problems in the distribution network such as reverse energy flows. Figure 3.2 shows the emerging view of energy flows once DG is introduced in the distribution network.

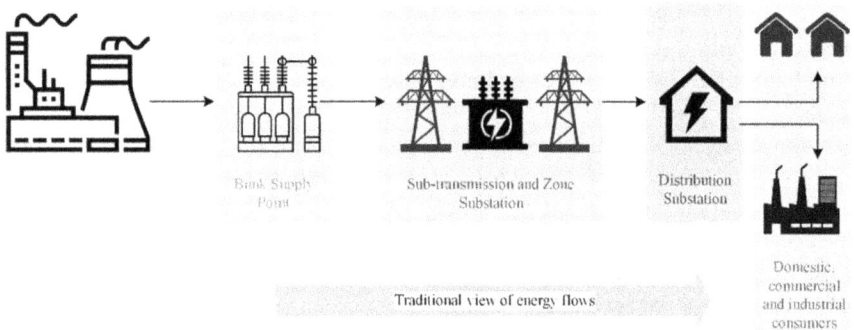

FIGURE 3.1   Traditional view of energy flows [1].

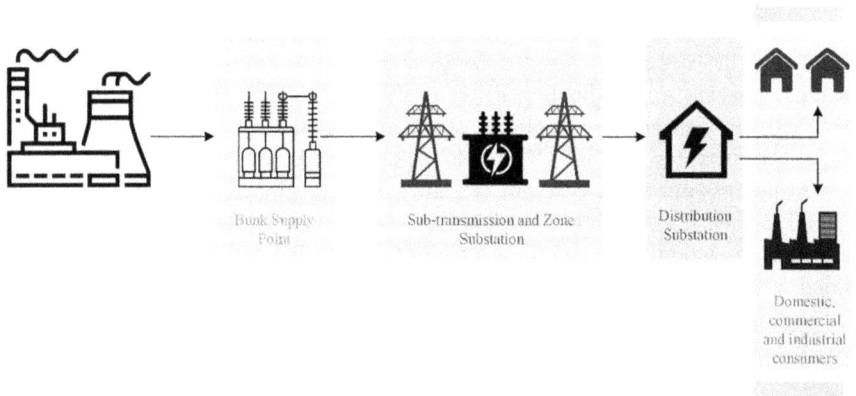

FIGURE 3.2   Emerging view of energy flows [1].

In this context, it is necessary to do a brief review of definitions, technologies, advantages, and disadvantages of this technology.

Definitions: Actually, there is not only one definition for DG. In the literature, a large number of terms and definitions are found to be associated with DG. The IEEE defines DG as the generation of electricity by facilities that are sufficiently smaller than central generating plants so as to allow interconnections at nearly any point in a power system [3]. The Electric Power Research Institute (EPRI) defines DG as generation from "a few kW up to MWs" [4]. The International Council on Large Electric Systems (CIGRE) defines DG as all generators with a maximum capacity between 50 and 100 MW connected to the electrical distribution system, which are not centrally designed or dispatched [5]. The Gas Research Institute contemplates that DG is "typically between 25 kW and 25 MW" [6]. From a constructional and technological point of view, DGs can be classified as shown in Figure 3.3. Each of these technologies is briefly described ahead.

3.2.2 STATE OF THE ART

This section presents the state of the art related to the impact of DG on electric power networks. In the current scientific literature, there are several published works concerning the integration of DG into distribution systems. However, the most representative works associated with each of the factors involved in formulating the problem and determining variables in the evaluation of the penetration of DG are described below.

Impacts of DG: Impacts of DG on electrical energy systems are inevitable, and therefore control requires the efforts of generation companies and users. The implementation of DG can be reflected in phenomena such as bus voltage, harmonics, power losses, reliability, etc. Through the inclusion of DG, the power flow can be bidirectional, and it can cause overvoltage in the distribution system. There are some interconnection guidelines in order to connect DG to the distribution network. In reference [7], rules for studying the impacts of interconnecting DG to a distribution feeder are defined. The IEEE 15474.7 standard [8] for distributed resource

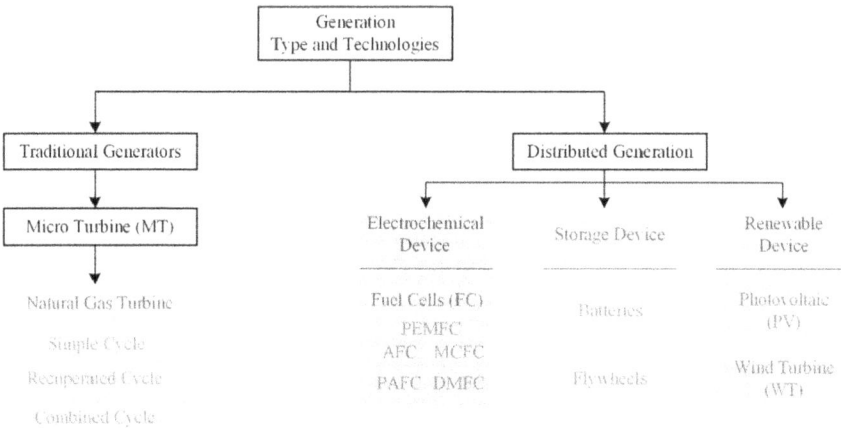

FIGURE 3.3 DG types and technologies [2].

interconnection provides technical criteria and requirements for interconnecting DG resources to distribution systems.

Voltage Impact: In scientific literature, there are many studies that have analyzed the impact of DG on technical parameters such as voltages, losses, capacity, reliability, and power quality. For example, in reference [9], the impacts of utility-scale PV-DG on power distribution systems are reviewed, particularly in terms of planning and operation in steady-state and dynamic conditions. One of the conclusions of this study is the improvement of voltage profiles when DG is considered. In this study, mitigation measures such as distributed storage are also presented with the purpose of reducing the magnitude of voltage fluctuation on the feeder analyzed [9]. In reference [10], a study of time series power flow analysis for distribution-connected PV generation is carried out in three real feeders and over the IEEE 8500 node feeder. The main purpose of this study was to find out how analysis using quasi-static time series (QSTS) simulation and high-time resolution data can quantify the impacts and the mitigation strategies to address voltage regulation operation and steady-state voltage. In reference [11], Chen et al. analyzed voltage problems due to DG penetration. The simulations were performed under the worst network condition (minimum demand and maximum DG output power). The Monte Carlo method was used to allocate DG. Also in this study, it was determined that a small amount of DG can produce voltage violations while very large amounts of DG allocated with adequate criterion do not affect the system operation. Navarro et al. [12] assess the impact of low-carbon technologies in low voltage (LV) distribution systems. Firstly, a realistic 5-minute time series daily profile is produced for photovoltaic panels, electric heat pumps (EHPs), electric vehicles, and micro-combined heat and power units. After that, a Monte Carlo simulation is carried out for 128 real UK LV feeders. The results of this study showed that PV panels technology produced problems in 47% of the feeders, EHPs produced problems in 53% of the feeders and EVs produced problems in 34% of the feeders. For uCHP, no problems were found. In reference [13], the impacts of DG on voltage regulation by load tap changing (LTC) transformers were studied. That study shows that if the LTC tap transformer control is not applied, problems of under-voltages and over-voltages can occur. In reference [14], energy storage is used for voltage support in low-voltage grids when high photovoltaic penetration is added. Centralized storage (CS) and distributed storage (DS) are investigated. CS is implemented by adding a single storage unit on a feeder node while DS integrated a storage device together with PV at a feeder location.

Impact on Electric Losses: In reference [15], the technical impacts of microgeneration on low-voltage distribution networks are analyzed. Impact analysis on losses demonstrates that an adequate amount of PV generation (30% PV) decreased the daily total losses in the system, while an inappropriate amount of PV (100%) generation increased losses. In reference [16], the impact of DG of a distribution network on voltage profile and energy losses are analyzed. The results obtained in this paper show that an adequate location and size of DG are essential for reducing power losses and voltage profiles. In reference [17], a probabilistic technique for optimal allocation of PV-based distributed generators to decrease system losses was developed. This research work presented six cases. For the base case, without DG, the annual energy loss is 540.967 MWh, while for case 6 when a DG of 1590 kW is added, the annual energy losses were reduced to 161.721 MWh, which represents energy and economic saving.

3.2.3   IEEE Standard 1547-7

In this section, we present a brief explanation of the main impact studies that should be developed according to IEEE standard 1547-7. The IEEE standard 1547-7 is a guide that describes criteria, scope, and extent for engineering studies of the impact on area electric power systems of a distributed resource (DR) or aggregate distributed resource interconnected to an area electric power distribution system [8].

In the following two tables, different conventional systems impact studies and their relationship to potential systems impacts on the electric power system (EPS) area are shown. In Section 1.2 of Chapter 1, it was established that the analysis of the distribution networks will be carried out in a steady state. Under this consideration, the green area painted in each one of the tables would correspond to the proposed analysis in that section. According to this standard, the steady-state study and the quasi-static study are briefly explained in this section.

Conventional Distribution Studies: According to this standard, a large DR could cause voltage variations, overload, equipment mis-operations, protection, and coordination, and power quality issues in the (EPS) area. Once DR is connected to the grid and if it exceeds some criteria limits, the DR should be studied using conventional studies which are presented in Table 3.1.

According to each of these studies, the impacts of the DR on the EPS area might be within acceptable limits. They also indicate that mitigations or more complex studies should be carried out.

TABLE 3.1
Type of Study vs Potential Impact on the EPS [17]

Impact	Study Type				
	Unintentional Islanding	Area EPS Equipment Duty and Operating Ratings	Protection Design, Coordination and Fault Rating	Voltage Regulation and Reactive Power Management	Power Quality
Steady-state simulation		✓	✓	✓	
System protection studies	✓	✓	✓		✓
Short-circuit analysis	✓	✓	✓		✓
Protective device coordination	✓	✓	✓		✓
Automatic restoration coordination	✓		✓		✓
Area EPS power system grounding			✓		✓
Synchronization			✓		✓
Unintentional islanding	✓		✓		✓
Arc flash hazard study			✓		
Operational characteristic loading, load shedding, etc.	✓	✓	✓	✓	

Steady-State Simulation: A steady-state simulation tool, also known as power flow simulation, solves a snapshot of an EPS model at nominal frequency. In its most basic form, this tool solves voltages, currents, real and reactive power flows, and losses throughout the area EPS at a single point in time. When these simulations are carried out, some types of problems can be identified from the results, such as:

- **Excessive Voltage Rise:** DR installation can cause a bus voltage to exceed acceptable limits due to the reverse flow of power caused during light loading conditions. Therefore, generation impact, such as solar, should be analyzed under the lowest daylight loading conditions. The concept of power flow study should include the low-voltage equivalent to understand the impacts on power quality at the equivalent customer terminals and to be able to analyze the performance of the DR connected to the low-voltage circuits.
- **Excessive Voltage Fluctuations:** DR, such as wind and solar generation, can cause voltage fluctuations which can be irritating to some customers. However, circuit voltage can be controlled through voltage regulators and capacitor banks.
- **Improper Operation:** When DR is added, it can create reverse power, which causes improper equipment operation. Voltage regulation equipment is designed to operate through power flow in only one direction from the source to the customer. Nevertheless, the inclusion of DS could cause a reversal of power flow from the customer to the source, and in this way, the equipment can wrongly adjust the voltage. This condition commonly happens under light load conditions.
- **Incorrect Situational Awareness**: If a large DR is installed on a circuit, circuit metering may be analyzed to know if the reverse power flow would be identified in readings provided to the system operators. Metering equipment should be replaced in order to capture bidirectional flow.
- **Equipment Overloads:** If the connection of DR is larger than the local load, an equipment overload can be caused in the area EPS. These overloads may occur at any time and not only at peak conditions.
- **Unbalanced Operation:** When the DR is installed at the location of area EPS with significant phase imbalances can cause voltage imbalance on the generator terminals. Single-phase DR installation can increase the imbalance, which in turn causes a serious impact on other devices connected to the area EPS circuits. Through a three-phase power flow analysis tool, modeling single-phase DR or three-phase DR on unbalanced circuits can recognize the complications that may occur on the circuit. Problems can be detected that occur in a circuit in which a single-phase DR or three-phase DR on an unbalanced circuit has been modeled.

According to this standard, "if the conventional steady-state simulation ("power flow" study) shows indications of equipment overloads, sustained overvoltage conditions, excessive voltage fluctuations, or equipment control problems, a quasi-static simulation should be considered to confirm the steady-state results or to analyze corrective measures" [17].

Special System Impact Studies: In many cases, special studies are not necessary. However, a need could arise to perform special studies even after the DR has been interconnected. For example, technical issues experienced after the DR interconnection, customer complaints, or a new DR application on the feeder might trigger some of these special studies. Table 3.2 shows the type of study and its relationship with potential impacts on the EPS.

TABLE 3.2
Type of Study vs Potential Impact on the EPS [17]

Impact	Study Type				
	Unintentional Islanding	Area EPS Equipment Duty and Operating Ratings	Protection Design, Coordination and Fault Rating	Voltage Regulation and Reactive Power	Power Quality
Quasi-static simulation	✓	✓	✓	✓	
Dynamic Simulation	✓		✓	✓	
Dynamic stability	✓		✓	✓	
System stability	✓		✓	✓	
Stability analysis interpretation	✓		✓	✓	
Voltage and frequency ride through		✓	✓	✓	
Electromagnetic transient simulation			✓		✓
Ferroresonanse			✓		
Interaction of different types of DR	✓	✓	✓	✓	✓
Temporary overvoltage		✓	✓	✓	✓
System grounding			✓		✓
DC injection		✓			✓
Harmonics and flicker		✓			✓
Harmonic analysis		✓			✓
Harmonic problems		✓			✓
Harmonic resonance		✓			✓
Flicker					✓

Quasi-Static Simulation: In this simulation, a sequence of steady-state power flow is conducted at a time step of no less than 1 second. However, it can be conducted using another time step, such as from 5 minutes to 1 hour. Applications such as energy and loss evaluation of generation and load profiles can be carried out under this solution mode. As is known, solar and wind energy are variable resources by a quasi-static simulation, and voltage fluctuation impacts due to variable DR output can be analyzed. The impact on voltage controls can be observed by quasi-static solution. Another advantage of this solution mode is that it can show the impact of DR on system equipment and customers.

3.3 INVERTER DESIGN

The inverter in photovoltaic applications is a crucial component of grid-connected PV systems as its converts the generated DC power from the panel into AC power which is supplied to the consumer. To separate the DC from the AC side, a transformer is used in the inverter to avoid the flow of leakage current from the AC to the DC side. However,, using the transformer makes the inverter bulky and introduces a lot of losses in the system resulting in a reduction of efficiency. With the advancement in inverter technology, the transformer has been replaced by the inverter which increases the efficiency and reduces the size of the inverter drastically. To limit the leakage, the filter design plays a vital role in transformerless inverters. The different classification of the inverters is presented in Figure 3.4.

PV inverters are classified according to the configuration as follows:

- Module-integrated inverters, typically in (> 500 kW)
- String inverters, typically in (1 – 10 kW)

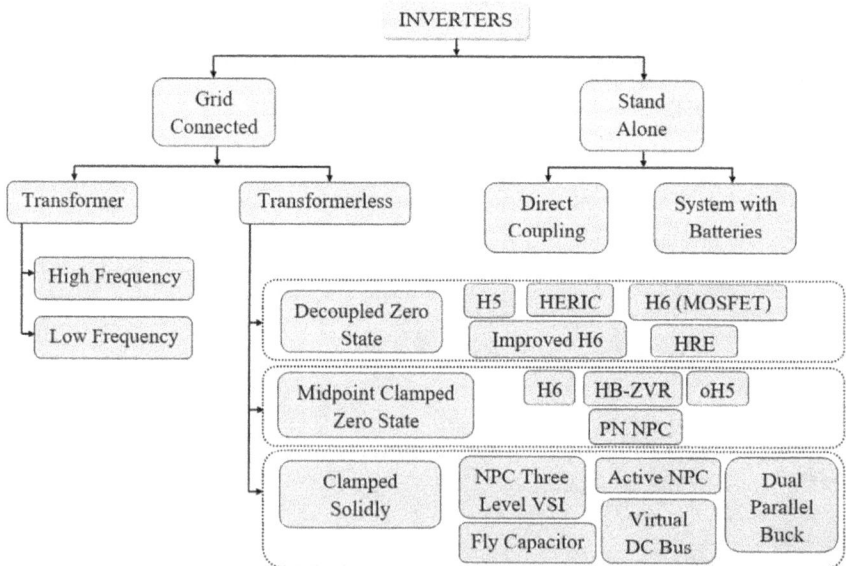

FIGURE 3.4 Classification of inverter topology.

- Multi-string inverters, typically in $(10 - 30 \ \text{kW})$
- Central inverters, typically in $(< 30 \ \text{kW})$

The above configuration are formed based on series and parallel combination of solar module and converters as per the system requirement. Module inverter comprises of converter for each panel and is also known as micro-inverters as depicted in Figure 3.5a. In a string inverter, a string of panels is connected to the converter as represented in Figure 3.5b. One more orientation in use is the multi-string inverter in which multiple strings are connected with the individual DC–DC converter which uses the MPPT to attain maximum efficiency from the panel and then connects it with a DC–AC converter

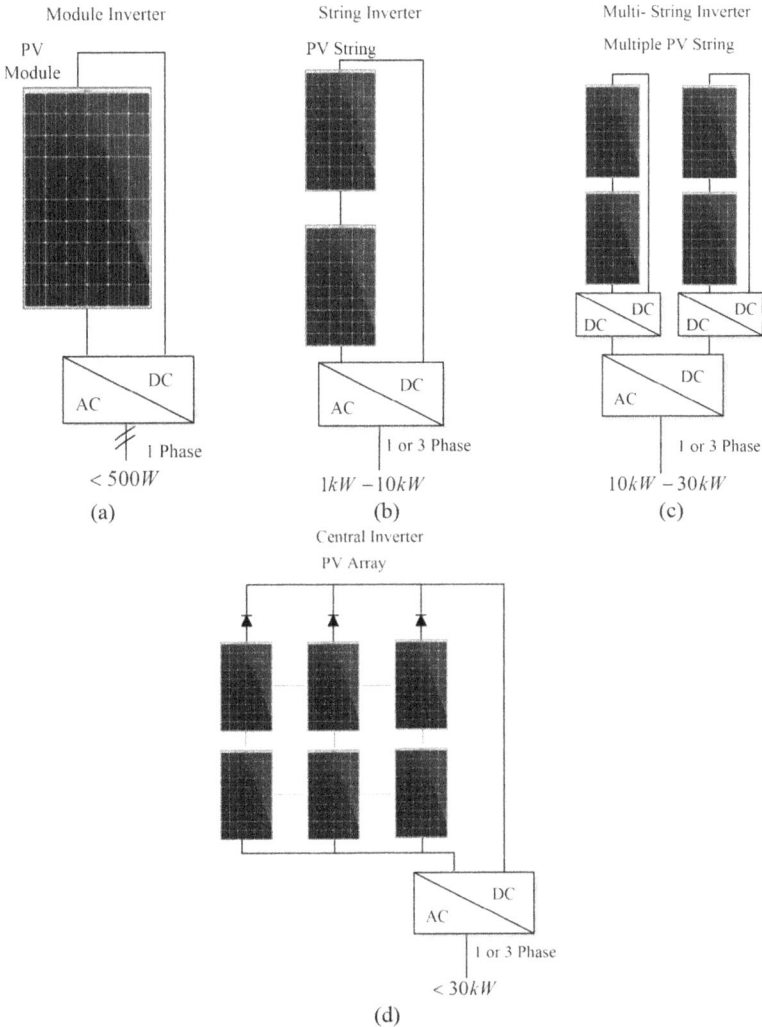

FIGURE 3.5   Configuration of inverter connections: (a) Module inverter. (b) String inverter. (c) Multi-string inverter. (d) Central inverter.

as shown in Figure 3.5c, whereas in the central inverter, multiple strings are connected to the single DC–AC converter as shown in Figure 3.5d. This connection is generally used for high-power generation. Mostly, the central inverter is used as it is low cost and high efficiency as losses are reduced compared to other methods.

Classification of grid-integrated PV transformerless inverter topologies can be carried out in the following three classes. The decoupling method and leakage current are used for the classification of transformerless topologies. The classes include various zero-state topologies like decouple, midpoint clamped, and clamped solidity. During freewheeling mode, decouple topologies can decouple the PV module from the grid. Moreover, the topologies that decouple the PV module from the grid during the freewheeling mode and in midpoint zero-state clamped topology short-circuit voltage clamping of the DC link from the midpoint have been included.

3.3.1 DECOUPLED ZERO-STATE TRANSFORMERLESS TOPOLOGY

The topology, in general, has a PV module connected to single-phase operation as a simple circuit when compared to half-bridge DC bus voltage. It is a low-cost topology, efficient, and a PV module connected in series. A few topologies under it are an H5 topology, a Highly Efficient and Reliable Inverter Concept (HERIC) topology, H6 metal–oxide–semiconductor field-effect transistor (MOSFET) topology, an improved H6 inverter topology, and an HRE topology [18].

3.3.1.1 H5 Transformerless Inverter

H5 is one of the most used transformerless inverter topologies. In this topology, a unipolar output voltage is generated by disconnecting the PV array from the load with the help of a power switch. This principle is widely utilized in the zero-voltage vector. In Figure 3.6, the principle of H5 topology is explained. The inverter operates in four modes with the help of five switches. During the active mode, three switches are in operation, whereas in the freewheeling condition, only two switches are in operation.

Modes 1 and 3: S1 and S4 switch operate at a switching frequency. The S1 switch is turned on continuously for the first half whereas S3 operates during the negative cycle. Between S4 and S5 high-frequency switching was carried out for obtaining a positive voltage vector. In the case of the negative vector for mode 3, high-frequency switching of S2 and S5 is performed. For the operation of mode 1, the S1 and S4 switch is turned on, and the flow of current is taking place from S5 to S1 and returns

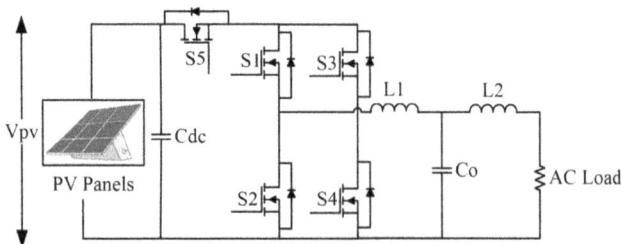

FIGURE 3.6   H5 transformerless inverter topology.

via S4, whereas in mode S3, the switch is turned on, and a current path is created from S5 to S3 and returns from S2.

Modes 2 and 4: In the freewheeling mode, S4 and S5 are off, and freewheeling takes place between the S1 and S3 switches.

3.3.1.2 Highly Efficient and Reliable Inverter Concept (HERIC) Topology

Among various transformerless inverter topologies, HERIC is regarded as one of the most critical topologies [21,22]. The HERIC topology is derived by adding an additional pair of antiparallel switches to the conventional H bridge inverter between switches S1–S4 and the filter. The topology is depicted in Figure 3.7. Additional branch is primarily provided to provide a third voltage level for the inverter which creates isolation between the AC and the DC side of the inverter. HERIC topology is most widely applied in transformerless topology.

Modes 1 and 3: Switching frequency is applied to S5 and S6 switches; S5 remains on continuously during the positive half of the cycle, and during the negative half of the cycle, S6 remains on continuously. A positive voltage vector can be obtained by switching S4 and S5 switches simultaneously at a very high frequency. The flow of current is held through S1–S5 when switch S5 is kept on, and it would return through S4. The flow of current is carried out via S3–S6 when the switch S6 is kept on, and it would return via S2.

Modes 2 and 4: Depending on the signal from the reference voltage S5 and S6 are turned on during the zero-voltage vector. PV is disengaged, and S1 and S4 remain off. A zero-voltage vector is achieved, and the unipolar output voltage is obtained, At the DC terminal, high-frequency input fluctuations are provided. During the freewheeling period, the inverter frequency is kept higher by shorting the load via S5 and S6 switches.

3.3.1.3 H6 (MOSFET) Topology

The H6 (MOSFET) transformerless inverter consists of six MOSFET and two freewheeling diodes (Figure 3.8). The topology presents a low leakage current because of the voltage which is applied through the capacitance forming a parasitic ground loop. The mode of operation is as follows.

Modes 1 and 3: During the positive half-cycle, the S4 remains on, whereas S1 and S6 operate in a switching frequency. When S1, S4, and S6 are on, the inductor

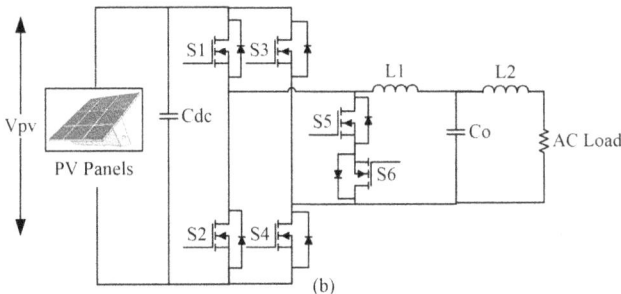

(b)

FIGURE 3.7 HERIC transformerless inverter topology.

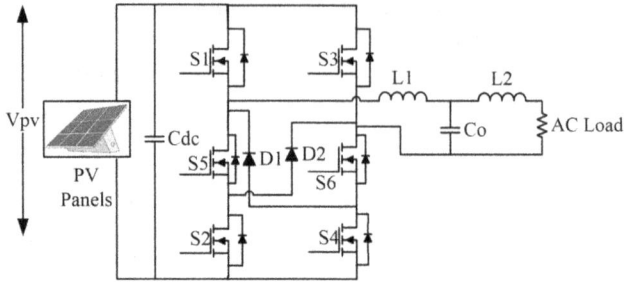

FIGURE 3.8   H6 (MOSFET) transformerless inverter topology.

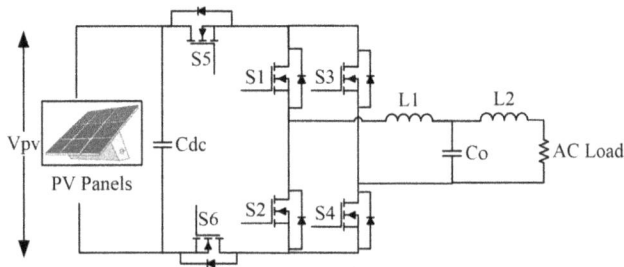

FIGURE 3.9   Improved H6 transformerless inverter topology.

current is charged. During the negative half-cycle, the S3 remains on and S1 and S5 operate in the switching frequency. The flow takes place from the S2 to the filter and then to S3 and S5.

Modes 2 and 4: Modes 2 and 4 are the freewheeling modes in which the positive cycle D1 is turned on to provide a freewheeling path from S4 to D1. After the negative cycle D2 is turned on, the freewheeling path is followed from S3 to D2. The remaining switches are turned off during this mode of operation.

3.3.1.4 Improved H6 Inverter Topology

H6 topology is very similar to that of the H5 topology. In the H6 topology, there is an extra switch S6 that is connected in series with the negative DC rail as depicted in Figure 3.9. The H6 topology operates like that of the S5 topology by completing the complete operation cycle in four modes of operation out of which two modes are freewheeling operations.

Modes 1 and 3: S1 & S6 Switches operate at a switching frequency. S1 and S6 switches are turned on continuously for the first half. Between S4 and S5, high-frequency switching was carried out for obtaining a positive voltage vector. In the case of negative, S2 and S6 operate at high frequency. For the operation of mode 1, S1 and S6 switches are turned on, and the flow of current takes place from S1 to S5 and returns via S4 and S6. A similar method is opted for the reverse path as well as an opposite pair of switches for mode 3 operation.

Modes 2 and 4: During zero-voltage switching, S1 and S4 switches are kept on, and the path for freewheeling is carried on. Moreover, opposite switches operate for reverse zero-voltage switching.

3.3.1.5   HRE Topology

The HRE topology comprises six MOSFETs and six diodes, and the AC coupling is split between two inductors. The AC switch pairs are present in a combination of S5–D5 and S6–D6 which provide a unidirectional current flow during the freewheeling phase which minimizes the leakage current.

Modes 1 and 3: During the positive half-cycle, the S5 remains on, whereas S1 and S3 operate in a switching frequency. When S1 and S3 are on and the D5 is in reverse biased, the inductor current is charged. During the negative half-cycle, the S6 remains on and, S2 and S4 operate at a higher frequency. The flow takes place from S2 to the filter and then to S4.

Modes 2 and 4: Mode 2 and 4 are the freewheeling modes in which the positive cycle S5–D5 combination is turned on to provide a freewheeling path from S5–D5 to L1. After the negative cycle S6–D6 is turned on, the freewheeling path is followed from S6–D6 to L2. The remaining switches are turned off during this mode of operation (Figure 3.10).

All the decoupled zero-state transformerless topologies are compared on the basis of different parameters in Table 3.3.

3.3.2   Midpoint Clamped Zero-State Transformerless Topologies

The midpoint clamped zero-state transformerless topologies are similar to decoupled zero-state topologies and most of the midpoint clamping topologies is obtained from the decoupling topologies. The major difference is that the output during the freewheeling mode is clamped to the midpoint of the DC bus.

3.3.2.1   H6 Inverter Topology

In this topology, two switches are added with the bidirectional clamping branch. The input is clamped by two diodes and two capacitors. The mode of operations is as follows (Figure 3.11).

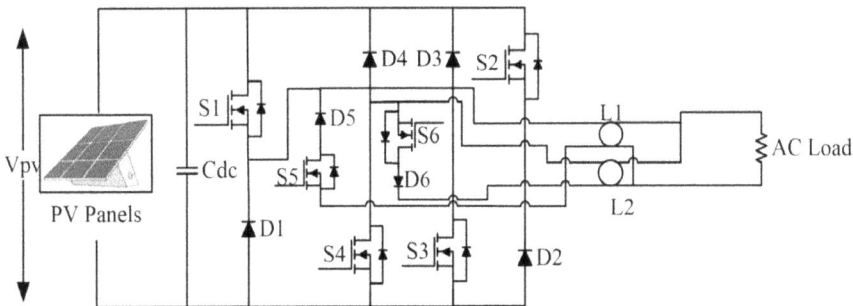

FIGURE 3.10   HRE transformerless inverter topology.

TABLE 3.3

Comparison Decoupled Zero-State Transformerless Topologies

Topologies			H6 (MOSFET)	Improved H6		
Parameters	H5	HERIC	Inverter	Inverter	HRE	Reference
Input capacitors	1	1	1	1	1	[19]
Input capacitance	Low	Low	Low	Low	Low	[20]
Switches in application	5	6	6	6	6	[21]
Diodes in application	0	0	2	0	6	[19]
Transistors voltage	1200	1200	250	600	600	[22]
Output voltage level	3	3	3	3	3	[18]
First harmonics	$2f_{sw}$	$2f_{sw}$	-	-	-	[20]
Electromagnetic interference	Low	Low	-	-	-	[22]
MPPTs present	1	1	1	1	1	[18]
Leakage current	Low	Low	Low	Very Low	Low	[23]
Maximum efficiency	98.5	-	98.3	97.1	99.3	[21]

Modes 1 and 3: Switches S1 and S4 operated during the positive half-cycle, whereas switches S2 and S3 operated during the negative half-cycle. The operation of the switches takes place at the line frequency.

Modes 2 and 4: Either diode D1 or diode D2 can be conducted in freewheeling mode, depending on whether the freewheeling path potential is higher or lower than half the intermediate circuit voltage. In this topology, the leak current elimination effect depends only on the switch-on speed of the clamp diode.

3.3.2.2 HB-ZVR Inverter Topology

During the no-load operation of HB-ZVR inverter topology, the short output voltage is limited to the midpoint of the DC bus through a diode rectifier and one bidirectional switch. The short, circuited output voltage during the freewheeling period is clamped to the midpoint of the DC bus through a diode rectifier and one bidirectional switch. To defend the lower DC-link capacitor from short-circuiting, an extra diode is connected as illustrated in Figure 3.12.

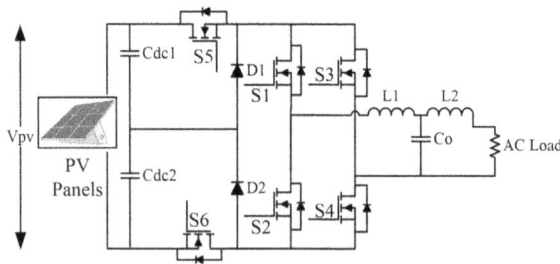

FIGURE 3.11   H6 transformerless inverter topology.

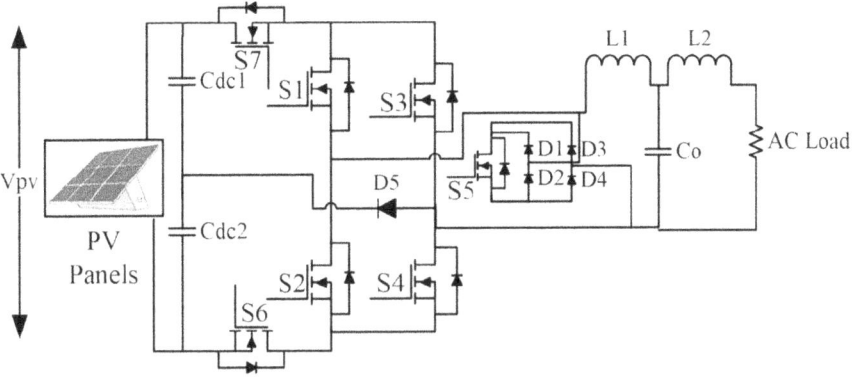

FIGURE 3.12   HB-ZVR transformerless inverter topology.

Modes 1 and 3: The positive half-wave of gate S5 is the opposite of gates S1 and S4, which have a small dead time to ignore short circuits in the network. During the negative half-wave, S5 is driven by opposite gate pulses from S2 and S3, shorting the inverter output and clamping it at the midpoint of the DC bus, creating a zero-voltage state.

Modes 2 and 4: The clamping function of this topology is realized with diode D5 which allows unidirectional clamping only if the potential of the free path is higher than the midpoint voltage of the DC link. As a result, the CM voltage can fluctuate when the opposite condition occurs.

3.3.2.3   H5 Type Inverter Topology

It is an advanced H5 topology in which a latching leg consisting of a switch and a capacitor divider is added to the H5 topology, clamping the freewheel potential to half the DC bus voltage as represented in Figure 3.13.

Modes 1 and 3: For the positive half-cycle, switches S1, S4, and S5 are operating, whereas for the negative half-cycle of operation, S2, S3, and S6 are operational.

Modes 2 and 4: For the freewheeling path to be fully restricted, S5 and S6 must be in a complementary transition, then S3 and D5 must be on for positive operation, and S5 and D3 must be on for negative freewheeling.

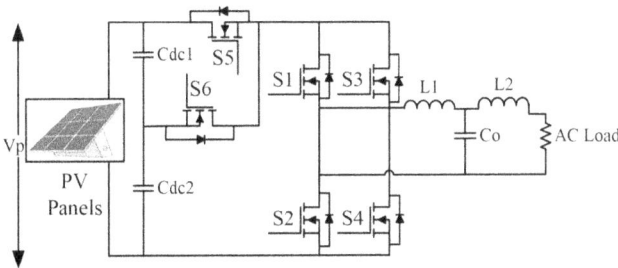

FIGURE 3.13   H5 transformerless inverter topology.

3.3.2.4 PN-NPC Inverter Topology

The neutral point clamping (NPC) topology is an excellent study of grid-connected PV systems. Two types of switching cells, i.e., positive and negative neutral cells, have been proposed for constructing the NPC topology as shown in Figure 3.14. During the freewheel period, the short-circuited output voltage is directly limited to half the DC input voltage via switches S7 and S8. Hence, the common-mode voltage remains constant $(V_{PV}/2)$ and the leakage current is low.

The different midpoint clamped zero-state transformerless inverter topologies are discussed in (Table 3.4).

3.3.3 CLAMPED SOLIDITY TRANSFORMERLESS TOPOLOGIES

In the clamped solidity transformerless topology, a stable connection is found between the grid and the PV module in freewheeling mode, so that the constant value of the high-frequency voltage CM tends to be constant.

3.3.3.1 Neutral Point Clamped Three-Level VSI Topology

The topology is most used for the high motor drive application where the switches are placed in one leg and diodes are used the clamp the middle point as illustrated in Figure 3.15. These clamp diodes guarantee a freewheel output current path in freewheel mode, this will result in an initial 0V. The topology is like the half-bridge with a higher efficiency and less ripple current. The common-mode voltage is present at a higher frequency so that the leakage current can be reduced. However, the requirement of 800V of input is a major drawback for the topology.

3.3.3.2 Active NPC Inverter Topology

The active NPC proposes a replacement of the diode with active power electronics switches. The switches as illustrated in the Figure 3.16 provide more clamping options. Hence, if switches S5 and S2 are on, turning on S3 and S6 sets the upper and lower clamp paths. The main function of the ANPC topology is to improve power distribution. As a result, the power semiconductor is charged evenly, and the efficiency of the converter is improved.

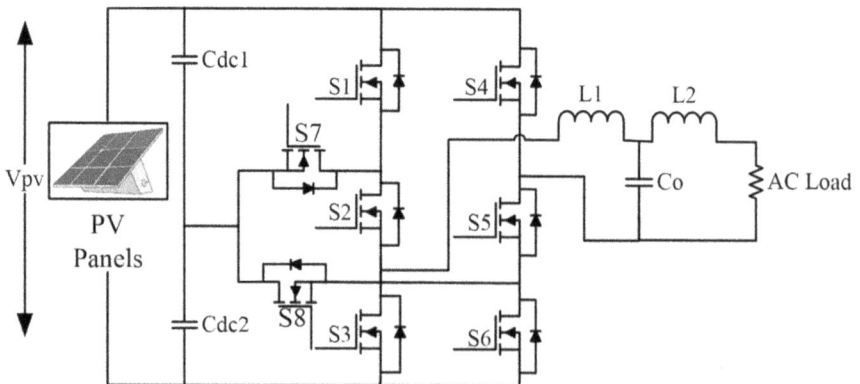

FIGURE 3.14 PN-NPC transformerless inverter topology.

TABLE 3.4
Comparison Midpoint Clamped Zero-State Transformerless Topologies

Topologies Parameters	H6 Inverter	HB-ZVR Inverter	oH5 Inverter	PN-NPC Inverter	Reference
Input capacitors	2	2	2	2	[24]
Input capacitance	High	High	High	High	[24]
switches in application	6	5	6	8	[21]
Diodes in application	2	5	0	0	[18]
Transistors voltage	1200	1200	1200	1200	[18]
Output voltage level	3	3	3	3	[24]
First harmonics	–	–	–	–	
Electromagnetic interference	Low	Low	–	–	[22]
MPPTs present	1	1	1	1	[20]
Leakage current	Very Low	Low	Very Low	Very Low	[25]
Maximum efficiency	97.4	94.88	-	97.8	[21]

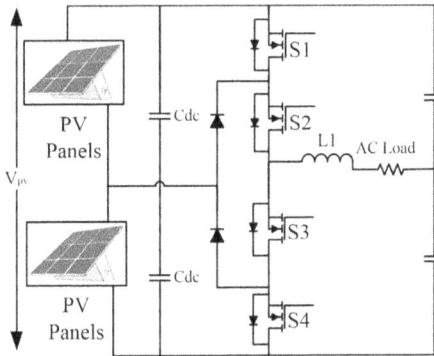

FIGURE 3.15   NPC 3 level VSI transformerless inverter topology.

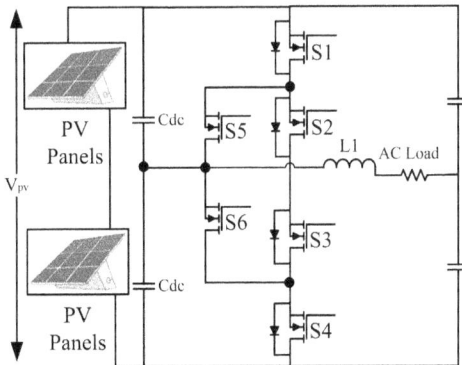

FIGURE 3.16   Active NPC transformerless inverter topology.

3.3.3.3 Dual-Parallel-Buck Converter Topology

It is a solid-clamped transformer topology as depicted in Figure 3.17. The objective of the derived topology is to attain reverse power flow. The negative output of the photovoltaic module is directly connected to the positive half-cycle inverter neutral and the negative half-cycle phase. As a result, high-frequency fluctuations in the common-mode voltage are minimized, resulting in lower leakage currents. In active mode, the inductor current flows through the two switches. Therefore, conduction losses are reduced. The major drawback of the topology is that if there is no dead time between switches S3 and S4 to which the network is directly connected, the network will be shorted.

3.3.3.4 Virtual DC Inverter Bus Topology

It is one of the most cost-effective PV inverter topologies. For reducing the leakage current the capacitor between the PV and the ground is connected to the neutral line of the grid directly with the negative pole of the DC bus as illustrated in Figure 3.18. The negative level is created by the DC bus. The inverter receives a three-phase output voltage just like the unipolar freewheeling diode inverter along with a good differential mode (DM) characteristic. In the positive half-cycle of the main current, switches S1 and S3 are always on, S2 is always off, and S4 and S5 complement each other with high frequencies. In a negative half-cycle, S5 is always on, S4 is always off, S1 and S3 are synchronous, and S2 additionally commutes to the switching frequency.

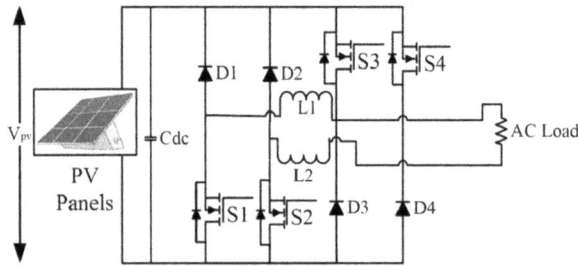

FIGURE 3.17 Dual-parallel-buck transformerless inverter topology.

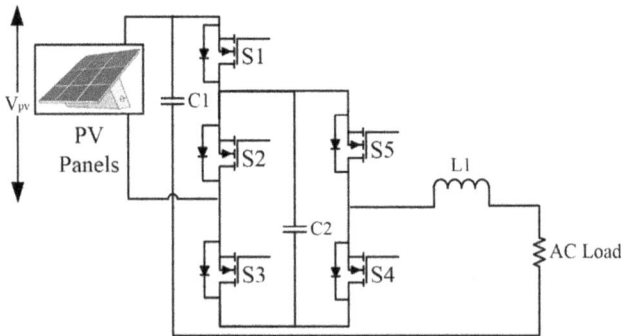

FIGURE 3.18 Virtual DC transformerless inverter topology.

3.3.3.5 Flying Capacitor Inverter Topology

The principle of operation for flying capacitor transformerless topology includes continuous charging and discharging of the capacitor connected between the branches for creating a negative supply voltage during the negative cycle of the inverter (Figure 3.19). When the positive end of the DC input is connected, then the flying capacitor is charged up to the input voltage provided by the PV array. Whereas when the negative half of the cycle DC side is switched to the capacitor, a negative voltage of equal magnitude appears at the output end. When the process is repeated continuously at a higher frequency, a constant output voltage appears across the load.

Modes 1 and 3: The S1 switch along with the D1 diode is kept on for the operation in the positive half-cycle. The capacitor is charged in this mode before its operation in the negative cycle to follow up. S2 is turned off for assisting the capacitor charging in this mode. Switching of the S3 switch is carried out at a very high frequency for obtaining a unipolar positive output voltage. A similar method is opted for the reverse path as well with an opposite pair of switches.

Modes 2 and 4: The S4 switch is turned on, and the S3 switch is turned off in this mode for obtaining zero voltage at the output. For the duration of the complete positive cycle, the capacitor remains connected to the input voltage. For an inverse application, opposite switches are in operation.

3.3.3.6 Conergy NPC Inverter Topology

Conergy NPC is one of the best alternatives to the classic NPC topology where the output voltage is clamped to the neutral point by using the bidirectional switch. In this topology, a zero-voltage state can be achieved by using switches S3 and S4 to clamp the output voltage to the ground (the midpoint of the DC bus). During the positive half-cycle, S1 and S3 complementarily commute with high frequency. And S2 and S4 complement each other with high frequency in the negative half-cycle. During the process, the common mode voltage is clamped to $V_{pv}/2$ which results in low leakage current flow through the capacitor (Figure 3.20 and Table 3.5).

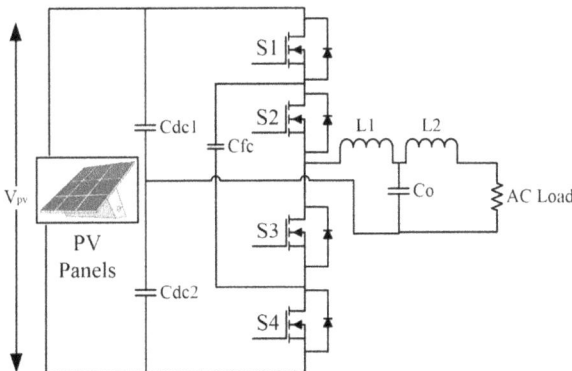

FIGURE 3.19 Flying capacitor transformerless inverter topology.

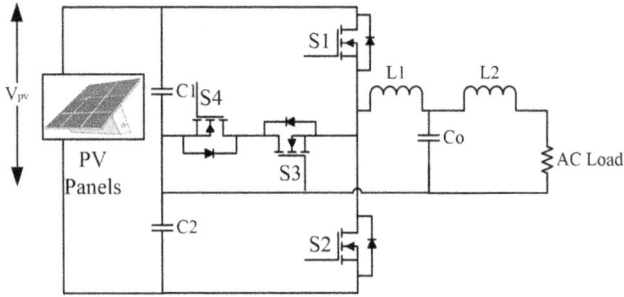

FIGURE 3.20 Conergy NPC transformerless inverter topology.

3.4 MODULATION TECHNIQUES

In this section, various pulse width modulation (PWM) techniques are discussed. In a voltage source inverter, it is essential to control the output voltage of the inverter. The output voltage of the inverter can be regulated by either considering AC voltage on the output end or DC voltage at the input end or internal control of the inverter. The most efficient method is by controlling the internal control of the inverter by implementing the PWM technique. In this method of control, a constant DC voltage is applied to the inverter and the output of the inverter is controlled by regulating the turn-on and turn-off period of inverter switches. The PWM inverters can be utilized in various industrial applications for DC to AC conversion and various voltage and frequency levels of AC load. Different types of modulation techniques can be classified as follows:

3.4.1 BIPOLAR PULSE WIDTH MODULATION

In the case of bipolar switching, antiparallel switches operate in complementary with other switches when operating at high frequency. For a basic H4 inverter topology, S1 and S4 are antiparallel with S2 and S3 switches. The zero value for output voltage cannot be achieved. The switching ripple present in the current is equal to the switching frequency resulting in larger filter requirements. Bipolar voltage is present across the filter resulting in high core loss, whereas the absence of common-mode voltage establishes that the leakage current present in the system is low. This PWM technique is not encouraged for transformerless inverters because of its low-efficiency performance (Figure 3.21).

3.4.2 UNIPOLAR PULSE WIDTH MODULATION

Both the legs of the basic H4 inverter are switched with sinusoidal reference to high frequency. A zero-output state is possible in two conditions: (a) S1 and S2 are on and (b) S3 and S4 are on. The switching ripple present in the current is twice that of the switching frequency, and as a result, the size of the filter is small. The voltage present across the filter is unipolar and hence the core loss is low. The presence of switching frequency components results in high leakage current because of which this type of modulation technique is not well suited for transformerless inverter topology (Figure 3.22).

TABLE 3.5
Comparison Clamped Solidity Transformerless Topologies

Topologies Parameters	NPC Three-Level VSI Inverter	Active NPC Inverter Topology	Dual-Parallel Buck Converter Topology	Virtual DC Bus Inverter Topology	Flying Capacitor Inverter Topology	Conergy NPC Inverter Topology	Reference
Input capacitors	2	2	1	2	3	2	[18]
Input capacitance	High	High	High	High	High	High	[21]
Switches in application	4	6	4	5	4	4	[20]
Diodes in application	2	0	4	0	0	0	[18]
Transistors voltage	600	1200	900	600	1200	1200	[18]
Output voltage level	3	3	3	3	3	3	[22]
First harmonics	$2f_{sw}$	$2f_{sw}$	$2f_{sw}$	$2f_{sw}$	$2f_{sw}$	$2f_{sw}$	[18]
Electromagnetic interference	Low	Low	Low	Low	Low	Low	[20]
MPPTs present	1	1	1	1	1	1	[21]
Leakage current	Very Low	Very Low	Low	Low	Very Low	Very Low	[26]
Maximum efficiency	98.16	97.34	99.0	–	–	97.67	[21]

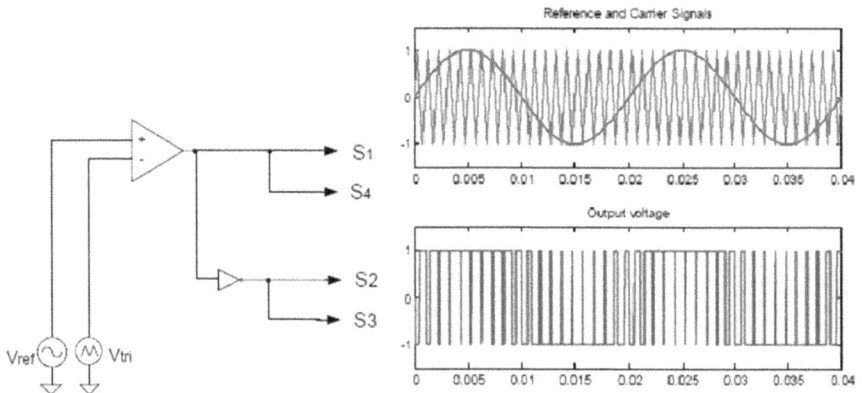

FIGURE 3.21   Bipolar PWM with 1 kHz triangular wave and 50 Hz sinusoidal reference.

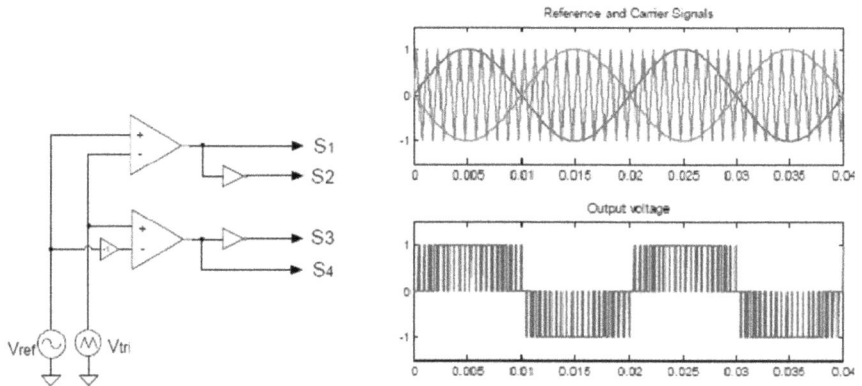

FIGURE 3.22   Unipolar PWM with 1 kHz triangular wave and 50 Hz sinusoidal reference.

3.4.3   Hybrid Pulse Width Modulation

In the hybrid modulation technique, one leg of the H4 inverter operates at grid frequency, whereas the other leg operates at a higher PWM frequency. The zero-crossing stage of the output voltage is present when S1 and S2 are on or S3 and S4 are on. The switching ripple in the current is equal to the switching frequency which results in a larger filter. The unipolar voltage present across the filter leads to low core loss. Also, the square wave variation at grid frequency causes high leakage current, which is undesirable for transformerless inverters (Figure 3.23) [27].

3.4.4   H5 (SMA) Modulation Technique

During zero-voltage, the grid is disconnected from the DC link by switching off the extra switch. High filtering is required as the switching ripple in the current is equal to the switching frequency. As the voltage across the filter is unipolar, the

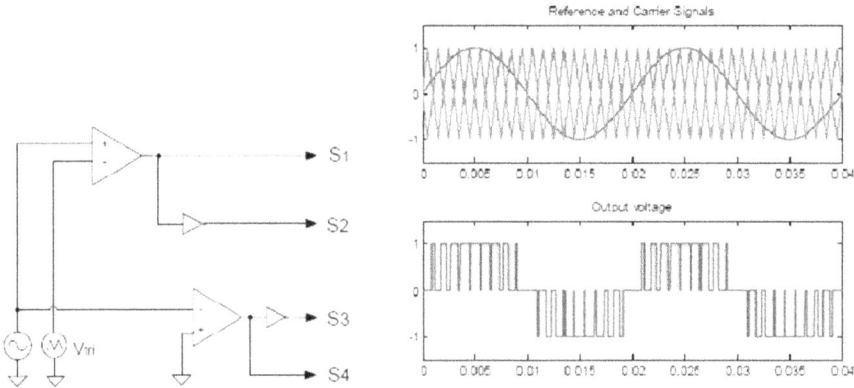

FIGURE 3.23 Hybrid PWM with 1 kHz triangular wave and 50 Hz sinusoidal reference.

core loss is low. The presence of a sinusoidal grid component reduces the leakage current of the techniques making it suitable for transformerless inverter application (Figure 3.24).

3.5 FILTER MODELING

For all bridge inverters, a low-pass output filter is needed to obtain the fundamental frequency output.

3.5.1 L FILTER

An L-type filter is shown in Figure 3.25. Over the frequency range, the attenuation range is − 20 *dB/dec*. A high-value inductor is required to attenuate the output current harmonics. A huge inductance means a bigger filter and more money. The system dynamics are harmed by the considerable voltage drop across the large inductor [28].

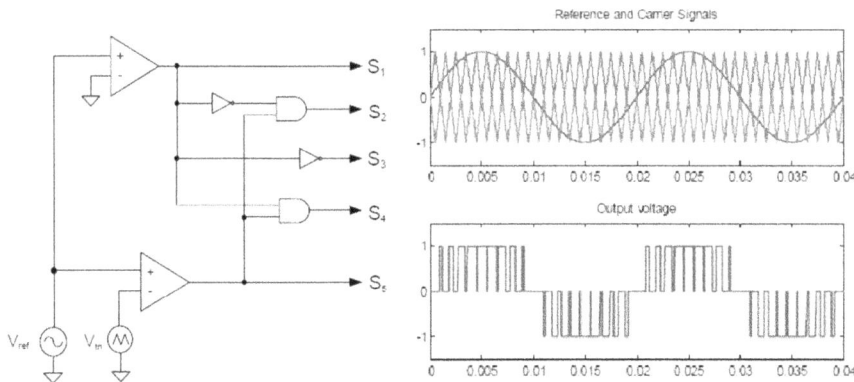

FIGURE 3.24 H5 SMA PWM with 1 kHz triangular wave and 50 Hz sinusoidal reference.

3.5.2 LC FILTER

The LC filter (Figure 3.26) is a second-order filter with a −40 dB/dec attenuation. The technique of creating an LC filter is rather simple. The design's trade-off is that a bigger capacitance may assist lower the inductor's cost. At the fundamental frequency, however, the system may experience inrush current and strong reactive current flow into the capacitor [28]. When an inverter is connected to the grid via an LC filter, the filter's resonance frequency is determined by the grid impedance [29]. Due to its small size and strong attenuation performance, the LC filter is a suitable choice for stand-alone inverters.

3.5.3 LCL FILTER

The third-order LCL filter, displayed in Figure 3.27, is widely used with both stand-alone and grid-connected inverters due to its high attenuation beyond resonance frequency. Compared to the other conventional filters, the LCL filter gives a better decoupling capability [28]. The design process of the LCL filter must consider the resonance of the filter and the current ripple flowing through the inductors.

The LCL filter design procedures are described and discussed in references [30,31]. According to reference [30], the formula used to calculate the minimum inverter-side filter inductance L_1 is given in equation 3.1.

$$L_1 > \frac{\varepsilon V_{DC}}{4 f_{sw} I_{rated} \xi} \tag{3.1}$$

When the inverter is unipolar modulated, the coefficient ε is 1. When the inverter uses bipolar modulation, the coefficient ε is 2. In equation 3.1, f_{sw} represents the switching frequency and I_{rated} is the inverter-rated current. g is the allowable current ripple ratio, and V_{DC} represents the DC input voltage. In this thesis, the allowed current

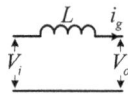

FIGURE 3.25   L-type filter.

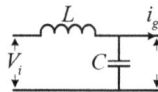

FIGURE 3.26   LC-type filter.

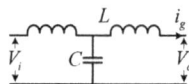

FIGURE 3.27   LCL-type filter.

ripple ratio ξ is chosen to be 15%. The maximum total inductance (L_1+L_2) should be less than the 10% system base inductance to avoid a large voltage drop across the inductors, as shown in equation 3.2. In equation 3.2, S_{rated} is the system-rated apparent power, and V_{rated} is the rated AC output voltage. Based on the calculation results from equations 3.1 and 3.2, the inverter-side inductance L_1 should be at least 1.9 mH, and the total inductance of the LCL filter should be less than 7.6 mH. As previously discussed, the capacitance is limited by equation 3.3. Using equation 3.4 to decide the grid-side inductance L_2. The coefficient γ is chosen to be 0.5 in equation 3.4.

$$L_1 + L_2 < 10\% \left(\frac{V_{rated}^2}{2\pi f_g V_{rated}^2} \right) \tag{3.2}$$

$$C < 20\% \left(\frac{S_{rated}}{2\pi f_g V_{rated}^2} \right) \tag{3.3}$$

$$L_2 = \gamma L_1 \tag{3.4}$$

Let the inverter output voltage and current be

$$v_0(t) = \sqrt{2} V_{0rms} \sin(\omega_0 t) \tag{3.5}$$

$$i_0(t) = \sqrt{2} I_{0rms} \sin(\omega_0 t + \phi) \tag{3.6}$$

Thus, the inverter instantaneous output power is

$$p_{out}(t) = v_0(t) \times i_0(t) = V_{0rms} I_{0rms} \cos(2\omega_0 t + \varphi) \tag{3.7}$$

In equation 3.7, there is a double-line frequency $(2\omega_0)$ component in the inverter output power. This double-line frequency harmonic is also seen in the inverter input. Thus, a DC-link capacitor should be utilized to limit the double grid line frequency voltage ripple at the inverter DC input. Sufficient capacitance would help to lower DC-link voltage fluctuations and reduce the inverter output current distortion, which is undesirable for power decoupling [32]. However, choosing an oversized DC-link capacitor increases design cost. The following analysis shows how to size the capacitance of a DC-link film capacitor for inverters.

Since the apparent power S_{rated} is

$$S_{rated} = V_{0rms} \times I_{0rms} \tag{3.8}$$

The output power is calculated as

$$p_{out}(t) = S_{rated} \cos\varphi + S_{rated} \cos(2\omega_0 t + \varphi) \tag{3.9}$$

The inverter input power is

$$p_{in}(t) = p_{out}(t) \tag{3.10}$$

Since the DC-link capacitor filters out high-frequency components, the double-line frequency component is expressed as

$$V_{dc}i_r(t) = S_{rated}\cos(2\omega_0 t + \varphi) \tag{3.11}$$

Thus, the double-line frequency current component at the DC side is

$$i_r(t) = \frac{S_{rated}}{V_{dc}}\cos(2\omega_0 t + \varphi) = I_r\cos(2\omega_0 t + \varphi) \tag{3.12}$$

Assume the dc side maximum ripple voltage to be 5% of the DC nominal voltage. The minimum DC-link capacitance is calculated as

$$C_f = I_r/2\omega_0 V_{rmax} = S_{rated}/4\pi f_0 V_{dc}V_{rmax} \tag{3.13}$$

Based on the analysis and calculations shown above, a minimum capacitance of 500 μF capacitor is needed.

3.6 SUMMARY

In this chapter, a detailed overview of the design and modeling of a grid-connected PV inverter along with filter design is established. The design aspects are based on a transformerless topology which is motivated at eliminating the drawbacks due to leakage current during the grid integration of DG system. The new topology has extra power electronics switches which are capable of providing freewheeling mode to the inverter during positive-to-negative cycle switching and vice versa. The presence of freewheeling mode reduces the flow of leakage current toward the DC side of the system. Furthermore, a brief discussion on the different transformerless inverter topologies, and available modulation techniques is provided and a comparison of the proposed topology with pre-existing topologies is presented. The analysis proved the efficiency of the developed topology in terms of components involved, harmonic levels eliminated, voltage regulation, and load balancing. Furthermore, parameters required for tuning the filter of the grid-connected converter in order to improve the performance of the converter are also discussed.

3.7 QUESTIONS

1. What is the concept of DG?
2. What are the types of DG?
3. What are the impacts of DG in the existing grid?
4. What is IEEE 1547.7 standard?
5. What are the problems reported due to distributed resources?

6. How is inverter topology classified?
7. What is the difference between string and multi-string inverters?
8. What is the difference between transformer and transformerless inverters?
9. Explain the working of H5 transformerless inverter topology.
10. Explain the working of HERIC transformerless inverter topology.
11. Explain the working of H6 transformerless inverter topology.
12. Explain the working of HRE transformerless inverter topology.
13. Explain the working of half-bridge ZVR transformerless topology.
14. Explain the working of neutral point clamp inverter topology.
15. Explain the working of neutral point clamped three-level VSI topology.
16. Explain the working of active NPC inverter clamp topology.
17. Explain the working of dual-parallel-buck converter topology.
18. Explain the working of virtual DC inverter buck topology.
19. Explain the working of flying capacitor working topology.
20. What are the modulation techniques used in the switching of inverter topologies?
21. Why is a filter required after an inverter in solar PV systems?
22. What is the difference between L-type, LC-type, and LCL-type filters?

REFERENCES

[1] D. W. Gao, *Energy Storage for Sustainable Microgrid*. Cambridge, MA: Elsevier, 2015.

[2] W. El-Khattam and M. M. Salama, "Distributed generation technologies, definitions and benefits," *Electr. Power Syst. Res.*, vol. 71, no. 2, pp. 119–128, 2004, doi: 10.1016/j.epsr.2004.01.006.

[3] V. S. Bhadoria, N. S. Pal, and V. Shrivastava, "A Review on distributed generation definitions and DG impacts on distribution system," *Int. Conf. Adv. Comput. Commun. Technol.*, November, pp. 1–7, 2013, doi: 10.13140/RG.2.1.4439.4328.

[4] O. Veneri, Ed., *Technologies and Applications for Smart Charging of Electric and Plug-in Hybrid Vehicles*. Cham: Springer International Publishing, 2017.

[5] K. Balamurugan, D. Srinivasan, and T. Reindl, "Impact of distributed generation on power distribution systems," *Energy Procedia*, vol. 25, pp. 93–100, 2012, doi: 10.1016/j.egypro.2012.07.013.

[6] P. J. Runci and J. J. Dooley, "Research and development trends for energy," in *Encyclopedia of Energy*. Amsterdam: Elsevier, pp. 443–449, 2004.

[7] Y. Baghzouz, "General rules for distributed generation - feeder interaction," in *2006 IEEE Power Engineering Society General Meeting*, 2006, p. 4, doi: 10.1109/PES.2006.1709238.

[8] IEEE, "1547.7–2013- IEEE Guide for Conducting Distribution Impact Studies for Distributed Resource Interconnection," 2014. doi: 10.1109/IEEESTD.2014.6748837.

[9] M. M. Begovic, I. Kim, D. Novosel, J. R. Aguero, and A. Rohatgi, "Integration of photovoltaic distributed generation in the power distribution grid," in *2012 45th Hawaii International Conference on System Sciences*, Jan. 2012, pp. 1977–1986, doi: 10.1109/HICSS.2012.335.

[10] R. J. Broderick, J. E. Quiroz, M. J. Reno, A. Ellis, J. W. Smith, and R. C. Dugan, "Time series power flow analysis for distribution connected PV generation," *Sandia Natl. Lab.*, no. January, p. 65, 2013.

[11] P.-C. Chen et al., "Analysis of voltage profile problems due to the penetration of distributed generation in low-voltage secondary distribution networks," *IEEE Trans. Power Deliv.*, vol. 27, no. 4, pp. 2020–2028, 2012, doi: 10.1109/TPWRD.2012.2209684.

[12] A. Navarro, L. F. Ochoa, and D. Randles, "Monte Carlo-based assessment of PV impacts on real UK low voltage networks," in *2013 IEEE Power & Energy Society General Meeting*, 2013, pp. 1–5, doi: 10.1109/PESMG.2013.6672620.

[13] C. Dai and Y. Baghzouz, "Impact of distributed generation on voltage regulation by LTC transformer," in *2004 11th International Conference on Harmonics and Quality of Power (IEEE Cat. No.04EX951)*, pp. 770–773, doi: 10.1109/ICHQP.2004.1409450.

[14] F. Marra, Y. T. Fawzy, T. Bulo, and B. Blazic, "Energy storage options for voltage support in low-voltage grids with high penetration of photovoltaic," in *2012 3rd IEEE PES Innovative Smart Grid Technologies Europe (ISGT Europe)*, Oct. 2012, pp. 1–7, doi: 10.1109/ISGTEurope.2012.6465690.

[15] P. B. Kitworawut, D. T. Azuatalam, and A. J. Collin, "An investigation into the technical impacts of microgeneration on UK-type LV distribution networks," in *2016 Australasian Universities Power Engineering Conference (AUPEC)*, Sep. 2016, pp. 1–5, doi: 10.1109/AUPEC.2016.7749321.

[16] V. Vita, T. Alimardan, and L. Ekonomou, "The impact of distributed generation in the distribution networks' voltage profile and energy losses," in *2015 IEEE European Modelling Symposium (EMS)*, Oct. 2015, pp. 260–265, doi: 10.1109/EMS.2015.46.

[17] E. A. Mohamed, Y. G. Hegazy, and M. M. Othman, "A novel probabilistic technique for optimal allocation of photovoltaic based distributed generators to decrease system losses," *Period. Polytech. Electr. Eng. Comput. Sci.*, vol. 60, no. 4, pp. 247–253, Sep. 2016, doi: 10.3311/PPee.10018.

[18] G. G. Iván Patrao, E. Figueres, F. González-Espín, "Transformerless topologies for grid-connected single-phase photovoltaic inverters," *Renew. Sustain. Energy Rev.*, vol. 15, no. 7, pp. 3423–3431, 2011, doi: 10.1016/j.rser.2011.03.034.

[19] T. Selmi, H. E. K. Baitie, and A. Masmoudi, "A novel single phase transformerless inverter topology for PV applications," in *International Conference on Sustainable Mobility Applications, Renewables and Technology (SMART)*, Kuwait, 2015, pp. 1–7.

[20] M. Islam, S. Mekhilef, and M. Hasan, "Single phase transformerless inverter topologies for grid-tied photovoltaic system: A review," *Renew. Sustain. Energy Rev.*, vol. 45, pp. 69–86, 2015, doi: 10.1016/j.rser.2015.01.009.

[21] M. Islam and S. Mekhilef, "Efficient transformerless MOSFET inverter for a grid-tied photovoltaic system," *IEEE Trans. Power Electron.*, vol. 31, no. 9, pp. 6305–6316, Sep. 2016, doi: 10.1109/TPEL.2015.2501022.

[22] E. Koutroulis and F. Blaabjerg, "Methodology for the optimal design of transformerless grid-connected PV inverters," *IET Power Electron.*, vol. 5, no. 8, p. 1491, 2012, doi: 10.1049/iet-pel.2012.0105.

[23] R. González, J. López, P. Sanchis, and L. Marroyo, "Transformerless inverter for single-phase photovoltaic systems," *IEEE Trans. Power Electron.*, vol. 22, no. 2, pp. 693–697, 2007, doi: 10.1109/TPEL.2007.892120.

[24] A. Das and G. Sheeja, "Photovoltaic H6-type transformerless inverter topology," in *2016 IEEE Annual India Conference (INDICON)*, Bangalore, 2016, pp. 1–6.

[25] D. Barater, E. Lorenzani, C. Concari, G. Franceschini, and G. Buticchi, "Recent advances in single-phase transformerless photovoltaic inverters," *IET Renew. Power Gener.*, vol. 10, no. 2, pp. 260–273, 2015, doi: 10.1049/iet-rpg.2015.0101.

[26] T. Kerekes, R. Teodorescu, and M. Liserre, "Common mode voltage in case of transformerless PV inverters connected to the grid," *IEEE Int. Symp. Ind. Electron.*, pp. 2390–2395, 2008, doi: 10.1109/ISIE.2008.4677236.

[27] R.-S. Lai and K. D. T. Ngo, "A PWM method for reduction of switching loss in a full-bridge inverter," *IEEE Trans. Power Electron.*, vol. 10, no. 3, pp. 326–332, May 1995, doi: 10.1109/63.387998.

[28] H. Cha and T.-K. Vu, "Comparative analysis of low-pass output filter for single-phase grid-connected Photovoltaic inverter," in *Applied Power Electronics Conference and Exposition (APEC)*, 2010, pp. 1659–1665.

[29] C. Lascu, L. Asiminoaei, I. Boldea, and F. Blaabjerg, "High performance current controller for selective harmonic compensation in active power filters," *IEEE Trans. Power Electron.*, vol. 22, no. 5, pp. 1826–1835, Sep. 2007, doi: 10.1109/TPEL.2007.904060.

[30] M. Bhardwaj, S. Choudhury, V. Xue, and B. Akin, "Online LCL filter compensation using embedded FRA," in *2014 IEEE Applied Power Electronics Conference and Exposition - APEC 2014*, Mar. 2014, pp. 3186–3191, doi: 10.1109/APEC.2014.6803761.

[31] J. M. Sosa, G. Escobar, P. R. Martinez-Rodriguez, G. Vazquez, M. A. Juarez, and M. Diosdado, "Comparative evaluation of L and LCL filters in transformerless grid tied converters for active power injection," in *2014 IEEE International Autumn Meeting on Power, Electronics and Computing (ROPEC)*, 2014, pp. 1–6, doi: 10.1109/ROPEC.2014.7036284.2014.

[32] X. Zong, *"A Single Phase Grid Connected DC/AC Inverter with Reactive Power Control for Residential PV Application,"* University of Toronto, 2011.

4 Standard for Operation of Distributed Generation Systems

4.1 INTRODUCTION

The objective of this chapter is to derive a medium-voltage grid code for distributed energy resource (DER) systems, to serve as a benchmark in assessing compliance of existing systems against anticipated grid requirements. It must be kept in mind that it is unrealistic to guarantee this single code will exactly model all future DER grid requirements; however, this is not the primary intention. Rather, the goal is to derive a grid code that best reflects future grid requirements based on (i) anticipated developments in North American grid requirements for DERs, (ii) dominant features and common trends of emerging grid codes, and (iii) requirements for a technically sound interconnection. The chapter is organized as follows: in Section 4.2, terminology commonly used by industry for categorizing the requirements within grid codes is explained. In Section 4.3, technical standards applicable to the grid integration of DERs within the distribution network are reviewed. In Section 4.4, a literature review on established wind energy system (WES) grid codes and emerging DER grid codes is conducted. In Section 4.5, a summary of limiting assumptions to assist in deriving a medium-voltage grid code is provided. In Section 4.6, the medium-voltage grid code is presented. In Section 4.7, the chapter summary is given.

4.2 GRID CODE TERMINOLOGY

The following terms are frequently used in grid code literature to characterize the performance and response capabilities of DER systems. A clear understanding of these terms and their related concepts is essential to correctly interpret specific grid code requirements. Therefore, as these terms are used extensively throughout subsequent sections, a brief explanation of each is provided.

4.2.1 POINT OF COMMON COUPLING

The point of common coupling (PCC) is defined by the IEEE as the point where a local electric power system is connected to an area electric power system [1–3]. This is not to be confused with the point of connection, which is the physical electrical connection between a DER unit and an electric power system. Figure 4.1 illustrates the relationships between PCC, point of connection, and an electric power system based on the IEEE convention [1–3]. Note the PCC and point of connection are not

DOI: 10.1201/9781003257189-4

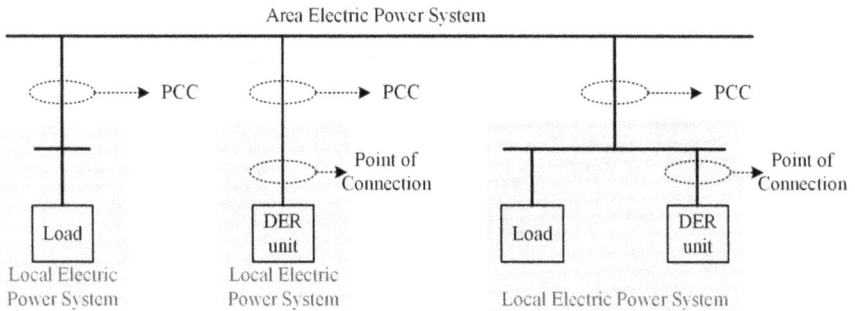

FIGURE 4.1   Definition of PCC and point of connection based on IEEE convention.

necessarily the same electrical point. Perhaps its most important feature, the PCC is almost exclusively classified by grid codes as the location where all system measurements are made and all interconnection requirements enforced.

4.2.2   AGREED ACTIVE POWER

This term denotes the maximum active power export permitted by a DER system for delivery into the distribution system and is contractually agreed upon by both the utility and DER system owner. It should be noted that there is no generally accepted definition for this term. For example, it is also referred to as the contract capacity [4] or project capacity [5–7]. The exact definition varies depending on the utility, network operator, and intended function of the DER system within the distribution system. Nonetheless, this quantity is typically the result of a connection impact assessment study performed prior to system installation. For DER systems with no local loads, the total nameplate capacity of all DER units is typically taken as the agreed active power. If including significant local loads, this may not be the case.

In this work, the agreed active power is defined as a fixed maximum amount of active power that can be injected into the distribution system by the DER system at the PCC. This limitation implicitly bounds the contractually obliged reactive power requirements of the DER system at the PCC. The agreed active power is determined after a connection impact assessment is performed and takes into account (i) the sum of the maximum active power outputs of each on-site DER unit, (ii) the aggregate load demands of the DER system, and (iii) intended DER system function(s) such as micro-grid capability or dedicated feed-in-tariff operation.

4.2.3   ANCILLARY SERVICES

These are the services provided by generators to support and ensure that the power system operates in a safe, secure, and reliable manner [8–10]. Traditionally, ancillary services have been supplied by synchronous generators and require a certain network reserve of active and reactive power regardless of load demands, to help maintain system frequency and voltage within expected limits. However, with the recent growth of DERs, the current model for providing network ancillary services is poised to change, to accommodate an active network involving bi-directional power flow from various de-centralized sources.

Ancillary services from DERs can provide advanced functions such as:

- Assist in supporting local voltage
- Assist in regulating network frequency
- Contribution to network spinning reserve
- Harmonic compensation
- Black start capability

Therefore, state-of-the-art power electronics provide a platform for DER systems to emulate many of the functions normally provided by centralized synchronous generators. In this work, ancillary services are defined as functions performed by the grid interfacing inverter(s), hence an overall DER system, which are beyond its fundamental task of transferring available active power from the DER system to the electrical network.

4.2.4 Voltage Sag and Swell

Within the context of FRT, many grid codes and technical standards loosely use various terms such as under-voltage, low-voltage, dips, and sags in order to quantify similar fault-induced drops in network voltage. Thus, to maintain consistency when discussing voltage deviations from steady-state values the term *voltage sag* is used throughout this work to indicate a momentary decrease in voltage magnitude from nominal (at fundamental frequency), regardless of fault duration and amplitude reduction. Conversely, the term *voltage swell* indicates a momentary increase in voltage magnitude from its nominal value. This convention is for the most part accurate with the deviations considered in this work for power system faults. However, it should be noted that there are actually several different sub-categories under which voltage deviations are classified depending on their relative magnitude, duration, and frequency content.

4.2.5 Fault-Ride-Through Requirements

These requirements specify an expected system response or behavior, both during and immediately following network fault conditions, where the DER system is prohibited from disconnecting. The enforcement of FRT requirements avoids two problems [11]:

- If a large capacity DER system is unable to ride through a fault, then it will disconnect from the network and introduce a sudden significant loss of generation that would normally not have occurred, possibly resulting in grid instability.
- The lack of voltage and frequency support for the network both during and immediately after the fault.

Depending on the network configuration and the type of fault (asymmetrical or symmetrical), a voltage sag within the vicinity of the fault propagates differently through network impedances [12,13], resulting in voltage sags at the DER system with varying degrees of severity. It is under these voltage conditions that grid interfacing inverters are required to ride through and thereby maintain the synchronous connection of the DER system. This imposed immunity to momentary voltage sags during grid faults is commonly referred to as low-voltage ride-through (LVRT).

4.2.5.1 Low-Voltage Ride-Through

LVRT requirements are defined by a voltage-versus-time curve similar to the example provided in Figure 4.2. When the PCC voltage is above the segmented regions in Figure 4.2, the DER system is required to maintain a synchronous grid connection. Conversely, when the voltage is within the segmented regions, the system is allowed to disconnect. Closely related to LVRT is the requirement to tolerate voltage swells. This imposed immunity to momentary voltage swells does not have a consistent designation in the literature but is sometimes referred to as high-voltage ride-through (HVRT).

4.2.5.2 High-Voltage Ride-Through

HVRT requirements are the complement to LVRT as they specify voltage swells that a DER system must remain synchronously connected during. Voltage swells occur less frequently than voltage sags during a grid fault. However, the presence of swells is still often associated with network fault conditions [14–16]. It is possible to define HVRT constraints by a voltage-versus-time curve, but it is most frequently characterized in the literature with a simple table of voltage ranges and corresponding trip times. The common causes for voltage swells include:

- A line-to-ground fault can increase the voltages on the healthy phases.
- Switching in capacitor banks onto the network.
- Sudden and significant reduction in system load (e.g., large loads switched off-line).

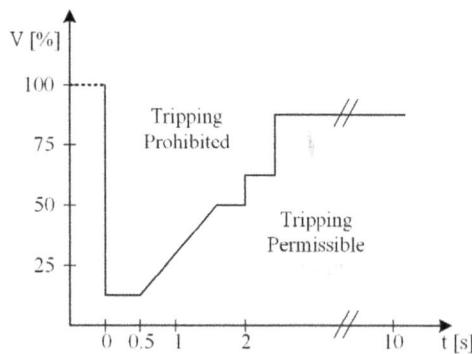

FIGURE 4.2 Sample LVRT curve.

4.3 TECHNICAL STANDARDS

Many technical bodies have released their own standards that outline technical requirements and practices for the interconnection of DERs. These standards encompass a broad set of topics such as power quality, response to network disturbances, and system operating limits. Utilities commonly use these standards as guidelines for developing their respective grid codes. The objective of this section is to identify the key standards governing the interconnection of DERs and to summarize their requirements corresponding to the assumptions of Section 4.5. The results will be used to develop a grid code that supports a technically sound interconnection for DERs. Due to the breadth of documentation available, the scope of standards is limited to North America.

4.3.1 IEEE STANDARD 929: IEEE RECOMMENDED PRACTICE FOR UTILITY INTERFACE OF PHOTOVOLTAIC SYSTEMS

IEEE Std. 929 [17] was released by the IEEE in 1988 to aid photovoltaic (PV) system designers, system owners, and utility operators with the interconnection of PV systems to a utility network. It has since gone through revisions, the latest being in 2000. The standard focuses on smaller capacity systems, 10 kW or less, which is characteristic of single-phase residential installations. Although in this work the main interest is in larger three-phase installations, intermediate (10–500 kW) and large-scale (over 500 kW) PV systems would likely utilize some modified form of the recommendations from this standard. This standard selects the PCC voltage operating range as a protection function rather than a voltage regulation function, as PV systems are considered to inject current into the utility network (due to converter control); not to regulate the line voltage. In the event of a voltage disturbance, maximum trip times are allotted for the system depending on the magnitude of voltage deviation. There are no explicit FRT requirements. This recommended PV system response to abnormal voltages is provided in Table 4.1.

TABLE 4.1
Response to Abnormal Voltages

Voltage[a] (at PCC)	Maximum Trip Time
$V < 60 (V < 50\%)$	6 cycles
$60 \leq V < 106 (50\% \leq V < 88\%)$	120 cycles
$106 \leq V \leq 132 (88\% \leq V \leq 110\%)$	normal operation
$132 < V < 165 (110\% < V < 137\%)$	120 cycles
$165 \leq V (137\% \leq V)$	2 cycles

[a] Voltage values are provided in both units of volts and a percentage of the base voltage of 120 V.

IEEE Std. 929 considers the frequency to be controlled by the utility and thus recommends a fixed frequency operating range of 59.3–60.5 Hz. For frequency deviations outside this range, the grid interfacing inverter is expected to disconnect from the network within six cycles (0.1 second). This nominal frequency operating window can be widened for small, isolated power systems, such as those on islands or in remote locations, as the system inertia would be lower for these cases. However, there is no mention of PV systems contributing to frequency regulation through active power control or any active power dependence on the frequency at all.

The PV system's recommended power factor operating range is ±0.85 when the output power is greater than 10% of the rated. IEEE Std. 929 assumes most systems are designed to inject power close to the unity power factor and thus have minimal reactive power abilities. No control schemes or strategies are recommended for controlling reactive power output to influence the local voltage. However, the standard recognizes that some systems may provide some form of reactive power compensation and therefore the power factor range may be widened based on approval from the network operator. The islanded operation of a PV system is prohibited in IEEE Std. 929. Once islanding conditions are detected, the system must cease to energize the utility network.

4.3.2   IEEE Std. 1547-IEEE Standard for Interconnecting Distributed Resources with Electric Power Systems

IEEE Std. 1547 [18] is well adopted by utilities and system manufacturers throughout North America as a technical guide for integrating DERs within the electrical network. The emergence of IEEE Std. 1547 in 2003 [19] was due to the increased and widespread growth of DERs, and the subsequent need for a set of standard technical requirements that focuses primarily on the interconnection of these DERs. Unlike IEEE Std. 929 [17], which targeted only PV systems, IEEE Std. 1547 was designed to be inclusive of all DG technologies. In conjunction with this standard, the application guide IEEE Std. 1547.2 [1] was released in 2008 to provide the detailed background and rationale of IEEE Std. 1547.

The scope of IEEE Std. 1547 covers installations based on any DG technology with an aggregate capacity of up to 10 MVA at the PCC. A lower capacity threshold of 30 kW is established to distinguish DER systems that have a minimal impact on power system operations [1]. The electrical networks under consideration for DERs interconnection include primary and/or secondary distribution voltage levels.

One of the most notable recommendations in IEEE Std. 1547 is that DER systems shall promptly cease to energize the distribution network once a fault is detected. There is no mention of FRT capabilities to aid in preserving grid stability, even though this standard provides recommendations for system capacities up to 10 MVA. This philosophy clearly implies that LVRT and HVRT are not endorsed. The basis for this practice is that when the distribution network detects a fault and subsequently de-energizes the affected circuit, then every other source on that circuit must also cease to energize it [1]. In other words, this lack of FRT expectation is based on an anti-islanding premise. The voltage and frequency operating ranges recommended by IEEE Std. 1547 are similar in nature to those of IEEE. Std 929. For abnormal voltages, the DER system must cease to energize the distribution network according to the clearing times in Table 4.2.

It is also recommended that DERs should not actively regulate the PCC voltage as this restriction will help prevent operating conflicts, such as interference with existing network voltage regulation practices (e.g., automatic line voltage regulation equipment) [20,21]. In fact, any network demand for reactive power support from a DER system is discouraged. For frequency disturbances, the DER system must disconnect within pre-specified clearing times as shown in Table 4.3.

This lack of imposed active power control is consistent with IEEE Std. 929. Unintentional islanding is not recommended in IEEE Std. 1547; however, this subject is under consideration for future revisions. Once islanding conditions are detected, the DER system must cease to energize the distribution network within 2 seconds. Very recent IEEE working group developments indicate that the IEEE 1547 series of standards are being revised to address the lack of FRT requirements [22,23]. In particular, two new standards are nearing completion. IEEE Stds. 1547.7 and 1547.8 are intended to fill the gap in IEEE Std. 1547 as a result of the high penetration of DERs, by supplying guidance for DER impact studies and providing supplemental support for implementation strategies to expand the usefulness of IEEE Std. 1547, respectively. Due to the widespread adoption of IEEE Std. 1547 by utilities and manufacturers, it is anticipated that these new standards will trigger a shift in the operating philosophy of DER systems toward one that preserves grid stability.

TABLE 4.2
Interconnection System Response to Abnormal Voltages

Voltage Range (% of base voltage[a])	Clearing Time (s)[b]
$V < 50$	0.16
$50 \leq V < 88$	2.00
$110 < V < 120$	1.00
$V \geq 120$	0.16

[a] Base voltages are the nominal system voltages stated in ANSI C84.1-1995.
[b] DR $\leq$ 30 kW, maximum clearing times; DR > 30 kW, default[2] clearing times.

TABLE 4.3
Interconnection System Response to Abnormal Frequencies

DR Size	Frequency Range (Hz)	Clearing Time(s)[a]
$\leq$ 30 kW	> 60.5	0.16
	< 59.3	0.16
> 30 kW	> 60.5	0.16
	< (59.8 – 57.0) (Adjustable set-point)	Adjustable 0.16–300
	< 57.0	0.16

4.3.3   CSA Std. C22.3 No.9-08-Canadian Electrical Code Part III: Interconnection of Distributed Resources and Electricity Supply Systems

The Canadian Standards Association (CSA) prepared Std. C22.3 No.9-08 [24] in 2008 to address the need for standardized connection practices. Its primary objective is to specify technical requirements for the interconnection of DER systems and electricity supply systems. As with IEEE Std. 1547, CSA Std. C22.3 No.9-08 covers the interconnection of DERs and distribution systems with voltages less than 50 kV and applies to any DG technology, with an aggregate system capacity of up to 10 MW at the PCC. This standard focuses on inverter-based systems connected to the distribution system where the PCC is at the medium-voltage level (defined as 0.75 kV through 50 kV). Contrary to the previously discussed standards, CSA Std. C22.3 No.9-08 allows the DER system to utilize voltage regulation practices at the PCC provided there is an agreement with the utility, and there is no detrimental impact on the PCC voltage levels. A power factor operating range of ±0.9 is recommended. For systems with less than 30 kW, the output power factor can be fixed, whereas systems that have an impact on distribution network voltages may be required to employ additional control strategies, such as dynamic power factor schemes.

CSA Std. C22.3 No.9-08 recommends DER systems implement under-voltage and over-voltage protection functions with maximum clearing times (i.e., no imposed FRT duration). These functions closely resemble the voltage requirements of IEEE Standards 929 and 1547. However, CSA Std. C22.3 No.9-08 allows LVRT should it be requested by the utility, in which case the set-point trip scheme outlined by the standard would no longer be valid; LVRT requirements would instead be decided upon after consultation between the DER system owner and utility. The voltage protective functions from CSA Std. C22.3 No.9-08 are provided in Table 4.4.

The under-frequency and over-frequency functions recommended by the CSA are similar to the requirements of IEEE Std. 929 and IEEE Std. 1547 but include an extended frequency operating range. There is no consideration for supporting network frequency through the control of active power. It is assumed the network regulates the frequency and the DER system disconnects once a nominal range is exceeded for a pre-specified time. These frequency functions are listed in Table 4.5 (DER capacity up to 30 kVA) and Table 4.6 (DER capacity above 30 kVA).

4.3.4   UL1741 Standard for Safety – Standard for Inverters, Converters, Controllers, and Interconnection System Equipment for Use with Distributed Energy Resources

UL1741 [25,26] is a well-established and commonly utilized standard that covers the construction, safe operation, and performance testing of inverters, converters, charge controllers, and interconnection system equipment associated with DERs. The following discussion focuses on the requirements associated with inverters and interconnection system equipment. In contrast to the previously discussed standards that provide technical recommendations, most utilities require inverters to obtain UL1741 certification as a prerequisite for grid connection.

TABLE 4.4
Response to Abnormal Voltage Levels

Voltage Condition at PCC, % of Nominal Voltage[a]	Clearing Time[b,c]
$V < 50$	Instantaneous – 0.16 s
$50 \leq V < 88$	Instantaneous – 2 s
$88 \leq V \leq 106$	Nominal operation
$106 < V \leq 110$	0.5 s – 2 min[d]
$110 < V \leq 120$	Instantaneous – 2 min
$120 < V < 137$	Instantaneous – 2 s
$137 \leq V$	Instantaneous

[a] Nominal system voltage shall be in accordance with Clause 6.2.
[b] Specific clearing times within the ranges in this table shall be specified by the wires owner. Other clearing times or voltage ranges may be arranged through consultation between the power producer and wires owner.
[c] Instantaneous means no intentional delay.
[d] Required for compliance with CSA CAN3-C235.

TABLE 4.5
Frequency Operating Limits for Distributed Resources ≤ 30 kVA

Adjustable Set-Point, Hz	Clearing Time (Adjustable Set-Point), s
59.3 – 57	0.1 – 2
60.7 – 57	0.1 – 2

Note: A fixed set-point can be acceptable in some jurisdictions; set-point should be confirmed with the wire's owner.

TABLE 4.6
Frequency Operating Limits for Distributed Resources > 30 kVA

Adjustable Set-Point, Hz	Clearing Time (Adjustable Set-Point), s
59.3 – 55.5	0.1 – 300
60.7 – 63.5	0.1 – 180

Note: More than one over-frequency and under-frequency set-point may be required by the wire's owner.

The vast majority of UL1741 content outlines the construction, performance, and testing of inverters; only a small portion discusses the interconnection requirements of these inverters with distribution networks. Within UL1741, the application of inverters is organized into two categories: (i) stand-alone and (ii) utility interactive. Stand-alone denotes an application that is not grid-connected, whereas utility-interactive implies an inverter that is connected in parallel with the electric utility.

As of January 2010, the latest revision of UL1741 requires that utility-interactive inverters or interconnection system equipment must comply with IEEE Std. 1547 and certain sections of IEEE Std. 1547.1 [19], the latter of which is a standard outlining standard conformance test procedures for equipment interconnecting DERs with electric power systems. Therefore, all recommendations summarized in Section 4.3.2 are valid for UL1741 as well. Because of this mandated compliance with IEEE Std. 1547, utility interconnection requirements are no longer explicitly provided in UL1741.

4.3.5   IEEE Std. 519 Recommended Practices and Requirements for Harmonic Control in Electrical Power Systems

IEEE Std. 519 [27] is widely adopted by many utilities as a baseline for assessing and dealing with harmonics injected into the power system by consumers [28]. This can be attributed to the fact that IEEE Std. 519 is endorsed by all the technical standards in Sections 4.3.1 through 4.3.4. IEEE Std. 519 applies to all types of static power converters interfaced with ac system voltage levels from 120 V through 161 kV, with an objective of establishing steady-state voltage and current waveform distortion limits.

IEEE Std. 519 defines a number of indices to be evaluated at the PCC in order to quantify the effect of harmonics on the power system. Based on these indices, harmonic limits are placed on the voltage and current waveforms. The indices relevant to the power quality criteria in this work are

- Voltage individual harmonic distortion (IHD)
- Voltage total harmonic distortion (THD)
- Current individual demand distortion (IDD)
- Current total demand distortion (TDD)

Voltage THD

$$THD = \frac{\sqrt{\sum_{h=2}^{H} V_h^2}}{V_1} \times 100\% \tag{4.1}$$

Here V_h is the rms value for voltage harmonic h, V_1 is the rms value of the fundamental voltage component, and H is the upper limit for the highest order harmonic being accounted for. It is important to mention that although the definition of H includes harmonic orders up to infinity, the generally accepted practice is to only include samples up to approximately $H = 50$. To calculate voltage IHD with respect to a particular harmonic h, equation 4.1 is used but with the numerator replaced by V_h.

Current TDD (Total Demand Distortion)

$$TDD = \frac{\sqrt{\sum_{h=2}^{H} I_h^2}}{I_L} \times 100\% \tag{4.2}$$

Here I_h is the rms value for current harmonic h, and I_L is the rms value of the maximum fundamental demand load (or generation) current. To calculate current IDD with respect to a particular harmonic h, equation 4.2 is used but with the numerator replaced by I_h.

Current and Voltage Distortion Limits
The current and voltage distortion limits relevant to a DER system are listed within IEEE Std. 519. The current distortion is restricted to a TDD of 5% at the PCC. The voltage distortion limits stipulate that for a distribution system voltage under 50 kV, the voltage IHD is limited to 3% of the fundamental while the voltage THD is limited to 5%.

4.4 GRID CODES LITERATURE REVIEW

4.4.1 WES SPECIFIC CODES

Due to the substantial growth and increased penetration of grid-connected WESs in recent years, many countries have instituted their own wind specific (or supplementary) grid codes, typically at the sub-transmission and transmission levels. These codes impose technical requirements governing the grid-connected operation of WESs. The primary reasons for the development of these grid codes were to [29]:

- Improve and stabilize wind turbine behavior.
- Reduce the amount of wind power to be lost following system disturbances.
- Provide WESs with operational characteristics similar to those of conventional synchronous generators.

An excellent discussion and comparison of existing WES grid codes across Europe and North America is available in [29,30]. Denmark and Germany were the first countries to have provisions within their existing grid codes for inclusion of WESs into their high-voltage networks [30] and are often considered leaders in the field, with other European nations having adapted their own grid codes accordingly. System operators in Canada and the USA have since followed suit and published their own WES grid codes similar to their European counterparts. In particular, the WES grid codes of Hydro-Québec [31] and the Alberta Electric System Operator [32] are often referred to in literature when referencing Canadian standards, as they are well-established and fairly progressive when compared to other provinces.

Each code has its own distinct set of requirements due to the respective power system's operating characteristics. Specific examples include (i) Germany, Spain, Great Britain, and Ireland require an injection of reactive current during grid fault events whereas other codes do not [29] and (ii) Hydro-Qu'ebec imposes an extended range of frequencies that the WES must ride-through due to a lack of synchronous links with neighboring networks [33]. However, although each WES grid code may have some unique properties, they all typically share a common set of requirements. These common requirements include:

- Voltage and frequency operating ranges
- Voltage regulation capabilities and associated reactive power control
- Frequency regulation capabilities and associated active power control
- FRT capabilities – namely LVRT and HVRT specifications
- Power quality metrics

It is also important to highlight the issue of contrasting (non-harmonized) WES grid codes as this is frequently discussed in the literature and has caused significant challenges for the wind industry in designing systems for compliance across multiple jurisdictions. This dissimilarity between codes is best illustrated with figures depicting LVRT and operating frequency requirements from several different WES grid codes, shown in Figures 4.3 and 4.4, respectively [29].

Although present WES grid codes may still experience revisions because of (i) increasing penetration levels, (ii) evolving needs of the power system, or (iii) efforts toward grid code harmonization, they have for the most part transitioned past the initial development and implementation phases. Most wind turbine manufacturers are now able to supply WESs that are compliant with these codes [29].

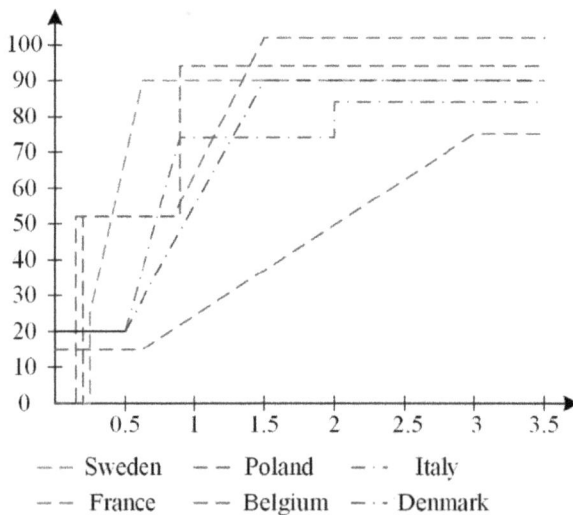

FIGURE 4.3   LVRT requirements of various WES grid codes.

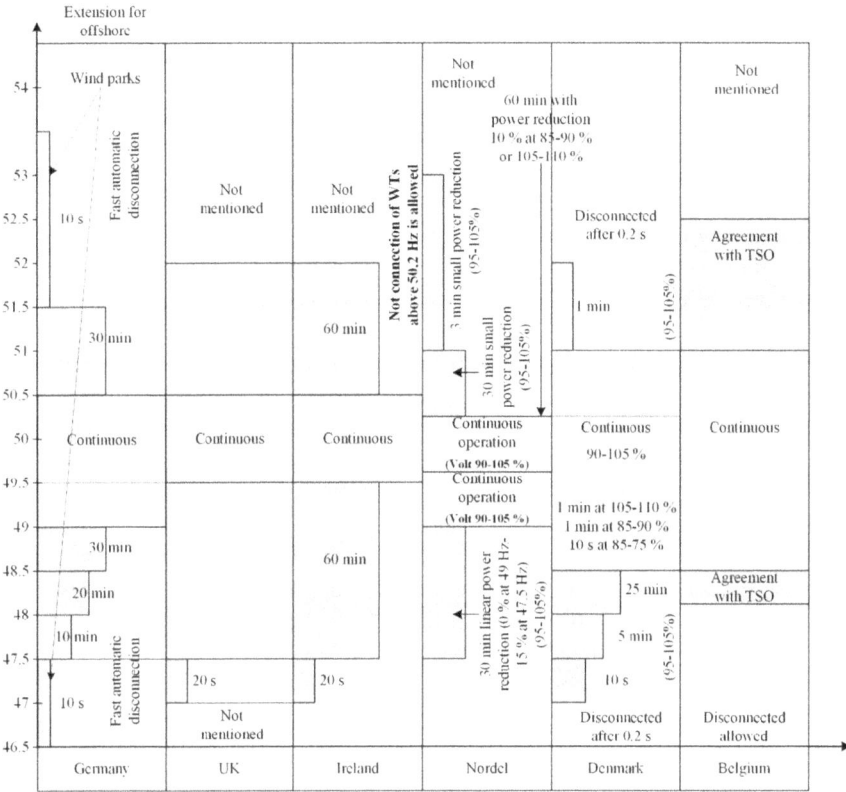

FIGURE 4.4 Operating frequency requirements of various WES grid codes.

4.4.2 DERs SPECIFIC CODES

Much like grid-connected WESs in recent years had evoked a re-design of subtransmission and transmission grid codes, the unprecedented growth of DG is contributing to the formulation of new medium-voltage grid codes governing the grid-connected operation of DERs. For the most part, these medium-voltage grid codes stipulate interconnection constraints for DER systems that are inclusive of all DG technologies. Due to the strong growth of DERs, particularly PV, Germany has comparably advanced technical rules for the network interconnection of DERs [34]. They are often referred to in literature with the release of their technical guideline for generating plants connected to their medium-voltage network [35–37] in 2008, which mandates both ancillary services and strict FRT requirements. A thorough discussion of this code is provided in reference [38].

This grid code is based on Germany's transmission network code [39] but adapted to meet the operating characteristics of a medium-voltage network. Germany has taken an active approach to incorporating DERs within their medium-voltage network, as network operators reserve the ability to request DER systems change their active and reactive power outputs to satisfy network conditions. Aside from Germany, other European countries are also publishing new (or revising existing) interconnection requirements to promote broader inclusion of DERs. Those following Germany's lead and enforcing codes with similar FRT requirements include Austria and France [40,41], with many others expected to follow suit shortly, such as Portugal [42], Spain, and Greece [43]. Outside of Europe, the picture is quite different. North American medium-voltage grid codes seem to be evolving slower and at present do not include stringent FRT requirements or advanced ancillary services. In 2010, Hydro-One released its own technical interconnection requirements for DG systems connecting to the medium-voltage network in the province of Ontario [5]. However, there are no expectations for FRT; a DG system is expected to simply disconnect within a pre-specified time once the PCC voltage deviates from a certain range. This lack of system FRT capability was taken directly from the recommendations in IEEE Std. 1547, which endorses an anti-islanding voltage protection scheme. Furthermore, there exist no provisions within the Hydro-One code for DG systems to incorporate frequency or voltage regulation abilities; active and reactive power control is mandated to adhere to basic operating limits. Even the Pacific Gas and Electric company, one of the largest electric utilities in the state of California involved with the highly ambitious California Solar Initiative Program [44], does not have an updated medium-voltage grid code. Introduced in 2005, their Electric Rule No.21 [45] is also consistent with the recommendations outlined by IEEE Std. 1547. Based on the preceding discussion it is anticipated that the existing medium-voltage grid codes of Hydro-One and Pacific Gas and Electric company are not sufficiently stringent to meet their respective government targets for DER integration. These codes are based heavily on IEEE Std. 1547, which was released in an era (2003) when power electronically interfaced DERs were never thought to reach the penetration levels being realized today and forecast for the future. As such, the medium-voltage grid code released in 2010 by Hydro-One should not be thought of as a modern document at the forefront of grid code development, but in fact, it should be appropriately perceived as an already outdated set of grid requirements that date back nearly a decade. The continued use of such codes that are designed for low penetration of DERs will result in transferring of costs away from DER owners and onto conventional generation producers and/or network operators.

It is encouraging, however, that although existing medium-voltage codes are less developed when compared to those in Europe, various organizations within North America are instituting task forces/working groups to assess the impact of large-scale DERs interconnected with the power grid. In particular, the North American Electric Reliability Corporation (NERC) initiated the integration of the variable generation task force (IVGTF) in 2007. In April 2009, the IVGTF released a preliminary report [46,47] which outlines several recommended actions for both the NERC and the industry. One notable action was to revise IEEE Std. 1547, to include voltage FRT performance requirements to reconcile with existing bulk power system requirements. In addition, the report recommended further evolution and reconciliation of IEEE Std. 1547 is needed to incorporate broader grid performance considerations, such as significant voltage and frequency excursions and voltage control.

In July 2011, an update on the IVGTF activities was provided in [48] where preliminary recommendations revealed an updated set of requirements that include provisions for LVRT. These IVGTF recommendations seem to be running in parallel with the developments outlined in Section 4.3.2, where the 1547 series of IEEE standards are under revision in order to adapt to increasing DER penetration levels. If implemented, these recommendations would have North American grid codes for DERs closer resembling their European counterparts.

4.5   LIMITING ASSUMPTIONS

To constrain the scope of the proposed medium-voltage grid code a set of limiting assumptions is provided. These assumptions are formulated to allow the code to outline a clear set of requirements focusing on key performance expectations of grid-connected systems, namely FRT capabilities and supply of ancillary services. The assumptions are as follows:

- The grid code is applicable to DER systems, connected at the medium-voltage (distribution) level. Systems shall be connected through the use of a dedicated interconnection transformer. The assumed distribution voltage Interconnecting a DER system through a dedicated transformer can eliminate many of the difficulties is ≤ 50 kV.
- The DER systems under consideration are three-phase installations with an agreed active power from 30 kW to 10 MW.
- All grid code requirements are evaluated at the PCC, which shall be taken as the high-voltage side of the interconnecting transformer located at the DER system site. This implicitly assumes any feeder cable connecting the DER system to the utility substation feeder is owned by the utility.
- The maximum active power export permitted by the DER system at the PCC shall be determined by the agreed active power as defined in Section 4.2.2.
- It is assumed a DER system will not have any operating limitations imposed on it due to site-specific network characteristics, such as the availability of distribution feeder capacity. Specifically, the distribution feeder's voltage regulation and fault protection schemes shall be assumed capable of supporting a high DER capacity.
- There are three main grid code categories: *Ancillary services*, *FRT requirements*, and *Power quality*.
- The category *Ancillary services* includes the functions associated with voltage and frequency support and associated active and reactive power control. All other secondary services are neglected. These requirements are enacted during steady-state network operation, which is defined to be when grid voltage and frequency are near their nominal values, and therefore do not apply during FRT.

The category FRT requirements shall satisfy three objectives:

- Prevent unwanted disconnection of the DER system. This is achieved by implementing both LVRT and HVRT requirements.
- Assist in stabilizing network voltage during faults. This is achieved by injecting a reactive current, which corresponds to an injection of reactive power.

- Assist in post-fault recovery of network frequency and voltage. This is achieved by appropriately controlling post-fault active and reactive powers.
- The category *Power quality* is limited to basic power quality constraints.
- DER systems, such as micro-grids, are prohibited from islanding with any portion of the distribution network. Once islanding conditions occur, isolation of the DER system from the network is required within 2 seconds. It is assumed the DER system has an appropriate anti-islanding detection and disconnection scheme which conforms to this requirement.
- Following disconnection of the DER system from the network, the system shall not reconnect until the network frequency and PCC voltage are within normal operating ranges, respectively, for a minimum of 5 minutes. It is assumed the DER system possesses a re-energization scheme that conforms to this requirement.
- Should the network operator or utility deem it necessary; it is assumed that the DER system has a readily available communication link to handle operator requests for ancillary services governing active/reactive power support.

4.6 MEDIUM-VOLTAGE GRID CODE

It is anticipated that as North American medium-voltage grid codes continue to evolve, they will (i) promote a more active role for DER systems within the distribution network, (ii) support network reliability and stability, and (iii) include operating practices and philosophies similar to those in European codes. In fact, such attributes are already endorsed by the recently issued CSA standard C22.3 No.9-08 with conditional allowance for LVRT and PCC voltage regulation practices. Moreover, both the NERC recommendations [46–48] and IEEE working group intentions [22] to revise the 1547 series of IEEE standards further reinforce that, as the installed capacity of DERs increases, sustaining a passive role for these DER systems within the distribution network will be inadequate.

Based on these anticipated developments a medium-voltage grid code applicable to this time constraint is based on the anti-islanding requirements of [19]. This time delay originates from [17].

DER systems are formulated into three main categories: Ancillary services, FRT requirements, Power quality.

The items within *FRT requirements* and *Ancillary services* are derived primarily from the requirements of the German medium-voltage grid code in [49]. However, modifications have been made; the origin of each modification will be specified as the code is presented. The entirety of *Power quality* is based on Section 4.3.

For clarity the following conventions are made:

1. Injection of active and reactive powers at the PCC are denoted by positive values of P and Q, respectively, as referenced in Figure 4.5.
2. The PCC current, I, assumes the direction indicated by Figure 4.5.
3. DER system voltage and output power requirements are stated as percentages.

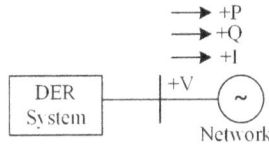

FIGURE 4.5   Power flow convention.

The PCC voltage, V, is expressed as a percentage of its nominal positive sequence value. For the most part, all output power (P, Q) requirements at the PCC are expressed as a percentage of the agreed active power or bounded reactive power. In all other cases, it will be explicitly stated as to what parameter the output powers are being expressed as a percentage of.

4.6.1   ACTIVE POWER CONTROL

4.6.1.1   Operating Range

The injection of active power at the PCC is limited to the DER system's agreed active power. The network operator may demand the DER system to either curtail its active power output or reduce active power output to zero, depending on network conditions. Accordingly, the DER system must be able to reduce its power output anywhere from 100% to 0% of the agreed active power, while at any operating point or from any operating condition. Reductions in active power output are either requested by the network operator or dictated by network frequency.

4.6.1.2   Manual Curtailment

The DER system must be capable of reducing its active power output to achieve a target set-point from the network operator. Depending on system capacity this target set-point can be achieved either directly or with decremental steps in output limited to 10% of the system's agreed active power. For any target set-point greater than or equal to 10% (of the system's agreed active power) the DER system must remain synchronously connected. Conversely, the DER system may disconnect for set-points less than 10%. The DER system shall achieve each target set-point from any initial active power output without intentional delay, and within a maximum time frame of 1 minute.

4.6.1.3   Automatic Curtailment

The DER system shall be capable of reducing its active power output corresponding to the over-frequency requirements outlined in Figure 4.6. Here f is the network frequency, ΔP is the required change in active power output expressed as a percentage of the DER system's active power export just prior to network over-frequency, the slope $d\Delta P/df$ represents the active power reduction rate, ΔP_{max} is the maximum required change in active power, and $(f_0, \Delta P_0)$ defines the curves origin. The DER system must be capable of curtailing its active power output by ΔP, according to the scheme in Figure 4.6, with a 97% settling time of no greater than 1 second. The frequency and active power parameters used in Figure 4.6 are summarized in Table 4.7.

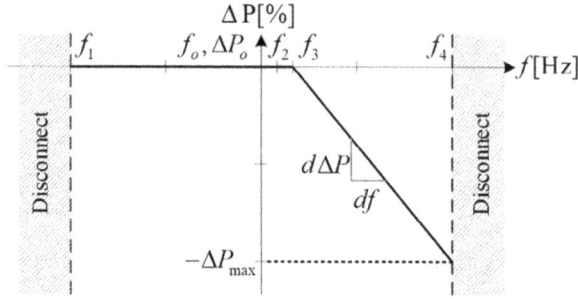

FIGURE 4.6 Medium-voltage grid code – Active power curtailment in response to network.

TABLE 4.7

Medium-Voltage Grid Code – Active Power Curtailment in Response to Network Over-Frequency: Parameters for Figure 4.6

Parameter	Value	Parameter	Value
f_0 [Hz]	60.0	f_4 [Hz]	60.5[a]
f_1 [Hz]	57.0[a]	ΔP_0 [%]	0
f_2 [Hz]	60.05	ΔP_{max} [%]	28
f_3 [Hz]	60.15	$\dfrac{d\Delta P}{df}$ [%/Hz]	80

[a] In keeping with IEEE Std. 1547 [19].

There are three distinct operating regions within Figure 4.6:

1. $f \leq f_1$ or $f \geq f_4$: DER system must immediately disconnect from the grid.
2. $f_1 < f \leq f_3$: DER system must remain connected, no imposed change in active power.
3. $f_3 < f \leq f_4$: DER system shall reduce active power output according to the curve in Figure 4.6, which can be expressed as

$$\Delta P = \frac{d\Delta P}{df}\left(f_3 - f\right) \tag{4.3}$$

It is important to stress that ΔP is expressed as a percentage of the DER system's active power output just before network over-frequency was detected $(f > f_3)$. By curtailing ΔP as a portion of the last measured active power export, as opposed to a portion of the agreed active power, the imprudent tripping of DER systems due to minimal active power export is avoided. For example, wind generators disconnect when their power output drops below a minimum threshold.

4.6.2 REACTIVE POWER CONTROL

4.6.2.1 Operating Range

During steady-state network conditions a DER system must be capable of supplying a reactive power output at the PCC corresponding to

$$0.9 \leq \cos\varphi \leq 1.0 \qquad (4.4)$$

Here $\cos\varphi$ is the displacement power factor at the PCC and φ is the angle between fundamental voltage and current waveforms, which can be positively or negatively valued. The displacement power factor is the ratio of active power to apparent power when considering only fundamental frequency voltage and current [50]. The displacement power factor is used in place of the true power factor as this is a reasonable approximation when meeting the harmonic requirements [49]. The power factor range in equation 4.4 must be satisfied across all possible active power outputs. Thus, the reactive power injection/ absorption at the PCC during steady state is bounded by the agreed active power coupled with equation 4.4. The DER system must be able to traverse this reactive power range within 1 minute, and whenever requested by the network operator or utility.

4.6.2.2 Static Voltage Support

The network operator may request a DER system to provide static voltage support during steady-state network conditions through reactive power control. There are two static voltage support schemes the DER system must be capable of providing:

- Reactive power output as a fixed value: Q = constant.
- Reactive power output to maintain a fixed power factor: .

Once a request is initiated for either scheme, the corresponding reactive power output must be achieved within 1 minute. The scheme enforcing a fixed power factor at the PCC translates to the family of straight lines shown in Figure 4.7, which is summarized as

$$Q = \left(\frac{0.9}{\sqrt{0.19}} \tan\phi \right) P, \text{ for } -\cos^{-1}(0.9) \leq \phi + \cos^{-1}(0.9) \qquad (4.5)$$

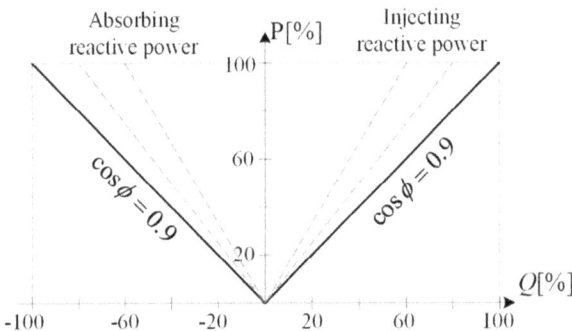

FIGURE 4.7 Medium-voltage grid code – Static voltage support: fixed power factor.

Here P is a percentage of the agreed active power and Q is a percentage of the maximum (absolute) valued reactive power output of the DER system, which is based on the agreed active power and the lower extreme of equation 4.4. No Q requirements are enforced for P less than zero as it is assumed the system always exports active power. These static voltage support schemes serve to prevent slow (local) voltage variations at the PCC from exceeding the steady-state voltage operating range. This is an important distinction from dynamic voltage support, which is initiated during sudden voltage disturbances at the PCC.

4.6.3 LVRT

The DER system shall adhere to the LVRT profile shown in Figure 4.8 for both symmetrical and asymmetrical grid faults. The voltage and time variables of Figure 4.8 signify they may take on different values at the network operators' discretion. Therefore, this voltage-time curve should be used as a guideline keeping in mind that minor curve variations are possible. However, that being said, Figure 4.8 does capture all the important features regarding emerging medium-voltage LVRT requirements. The specific voltage and time values assumed for Figure 4.8 are shown in Table 4.8.

In Figure 4.8, there are two distinct piece-wise linear curves. The DER system must ride through voltage sags resulting in a positive sequence fundamental voltage at the PCC that remains above Line 1. The DER system may disconnect for voltages

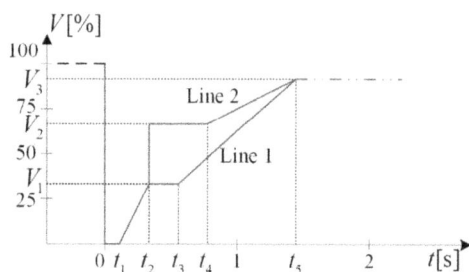

FIGURE 4.8   Medium-voltage grid code – LVRT requirements.

TABLE 4.8
Medium-Voltage Grid Code – LVRT Requirements: Parameters for Figure 4.8

Parameter	Value	Parameter	Value
t_1 [s] cycles	0.125 (7.5)	t_5 [s] cycles	1.25 (75)
t_2 [s] cycles	0.250 (15)	V_1 [%]	30
t_3 [s] cycles	0.442 (26.5)	V_2 [%]	70
t_4 [s] cycles	0.583 (35)	V_3 [%]	90

falling below Line 1. The employed voltage measurement scheme must be immune to the 120 Hz voltage component appearing in V due to negative sequence voltages arising from unbalanced faults. Note that Figure 4.8 requires the DER system ride through a fault resulting in a PCC voltage of 0% for 125 ms, and tolerate indefinitely a PCC voltage of 90%. For any voltage remaining above Line 2 a capacitive $(+Q)$ reactive current is to be injected by the DER system. This dynamic voltage support scheme is outlined in Section 4.6.5. Should the PCC voltage remain between Line 1 and Line 2, reactive current injection is not required but the DER system must still ride through the fault. Should the DER system remain connected following a voltage sag, the DER system's post-fault: (i) absorption of reactive power shall not exceed its pre-fault absorption amount, and (ii) active power output shall be restored to its pre-fault value with a gradient no less than 20% of the system's agreed active power per second.

4.6.4   HVRT

The DER system shall adhere to the HVRT requirements listed in Table 4.9 and thus remain synchronously connected for the minimum trip times indicated. During these voltage swells, an inductive $(-Q)$ reactive current is to be injected into the network by the DER system according to the dynamic voltage support scheme described in Section 4.6.5. Should the DER system remain connected following a voltage swell, the DER system's post-fault: (i) injection of reactive power shall not exceed its pre-fault injection amount, and (ii) active power output shall be restored to its pre-fault value with a gradient no less than 20% of the system's agreed active power per second.

4.6.5   DYNAMIC VOLTAGE SUPPORT

The dynamic voltage support scheme to be implemented during LVRT and HVRT conditions is shown in Figure 4.9a. Here In is the DER system's nominal current based on rated system kVA, V_n corresponds to the nominal positive sequence PCC voltage, ΔI_q is the relative amount of capacitive $(+Q)$ or inductive $(-Q)$ reactive

---

TABLE 4.9
Medium-Voltage Grid Code – HVRT Requirements

Voltage, V [%]	Minimum Trip Time, t [s] (cycles)
$100 \leq V < 110$	Unlimited
$110 \leq V < 120$	1 (60)
$120 \leq V < 125^a$	0.167 (10)
$125 \leq V$	Instantaneous

[a] These HVRT requirements are based primarily on [19]; however, this upper limit was selected from the HVRT constraints in [51].

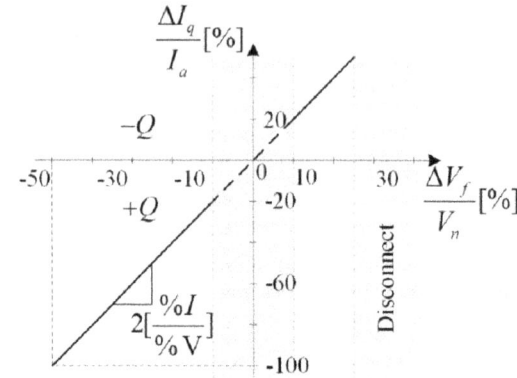

(a)

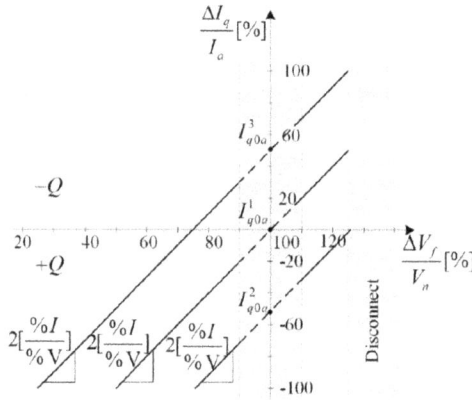

(b)

FIGURE 4.9 Medium-voltage grid code – Dynamic voltage support: (a) reactive current injection scheme and (b) reactive current injection examples.

current injected by the DER system during the fault, and ΔV_f is the relative change in PCC voltage during the fault. The parameters ΔI_q and ΔV_f are defined as

$$\Delta I_q = I_q - I_{q0} \tag{4.6}$$

$$\Delta V_f = V_f - V_n \tag{4.7}$$

where I_q is the instantaneous DER system reactive current injected during the fault, I_{q0} is the DER system reactive current output just prior to the fault, and V_f is the positive sequence PCC voltage during the fault. The required I_q shall be injected at the low-voltage side of the dedicated interconnection transformer rather than at the PCC. Note, reactive current is defined here as fundamental frequency current that lags or leads the fundamental frequency voltage at the point of injection by 90°.

In Figure 4.9a, a $\pm 10\%$ voltage dead-band disables any initial reactive current injection. Once the PCC voltage is detected outside this dead-band, the control system must initiate this scheme within one cycle (16.7 ms). For a severe voltage sag or swell, the DER system must be capable of supplying an $|I_q|$ corresponding to 100% of In. Because of the wide range of possible pre-fault currents, the condition of having $|I_q| = I_n$ is satisfied by a range of PCC voltage levels both above and below nominal. Based on this observation, a $\Delta I_q = I_n$ less than 100% may be required to achieve $|I_q| = I_n$. In any event, when the PCC voltage returns to the dead-band region, reactive current injection must continue for an additional 500 ms according to the curve in Figure 4.9a, to further aid voltage recovery. To illustrate the dynamic voltage support scheme of Figure 4.9a, three example reactive current injections are shown in Figure 4.9b. Here I_{q0}^1, I_{q0}^2, and I_{q0}^3 are pre-fault reactive current outputs of the DER system corresponding to 0%, -50% $(+Q)$ and $+50\%$ $(-Q)$ of I_n, respectively. Each example curve, with its respective I_{q0}, shows both LVRT and HVRT trajectories.

In the event of a voltage sag, capacitive reactive current is injected if the PCC voltage remains above Line 2 in Figure 4.8. As Figure 4.9b shows, the PCC voltage must drop to a sufficiently low value for $|I_q|$ to equal I_n. Moreover, a larger PCC voltage drop is required to have $|I_q|$ saturate at I_n when the pre-fault reactive current is more inductive. In the event of a voltage swell, Figure 4.9b shows that inductive reactive current is required to be injected until the PCC voltage reaches a maximum of 125% (due to voltage constraint in Table 4.9), regardless of the value of pre-fault current. Therefore, a greater amount of reactive power is absorbed by the DER system during HVRT for pre-fault currents that are more inductive.

4.6.6 POWER QUALITY

Harmonics: The DER system shall follow the requirements of [27] for permissible current and voltage distortion at the PCC. A summary of these harmonic current and voltage distortion limits is provided in Table 4.10. THD and TDD are defined in equations 4.1 and 4.2, respectively.

TABLE 4.10
Medium-Voltage Grid Code – Harmonic Current and Voltage Distortion Limits

Maximum Harmonic Current Distortion

IDD – Odd Harmonics[a] (%)

$h < 11$	$11 \le h < 17$	$17 \le h < 23$	$23 \le h < 35$	$35 \le h$	TDD
4.0	2.0	1.5	0.6	3.0	5.0

Maximum Harmonic Voltage Distortion

IHD (%)	THD (%)
3.0	5.0

[a] IDD for even harmonics are restricted to 25% of the odd harmonics limits.

Voltage: Under normal operating conditions the DER system shall not cause the PCC voltage to exceed ±6% of the nominal voltage rating. This requirement is based on standard CSACAN-3-C235–83 [52], which establishes preferred voltage levels for ac systems (from 0 to 50 kV) and is frequently adopted by utilities within their distribution grid codes [5,53]. The purpose of this voltage requirement is to facilitate proper sizing of DER systems relative to distribution feeder capacity at the design stage and is not related to any voltage ranges imposed by FRT requirements for LVRT and HVRT.

4.7 SUMMARY

This chapter developed a medium-voltage grid code based on anticipated grid requirements for DERs. This code was derived considering a number of driving factors. These factors include (i) recent grid code advancements in Europe, (ii) an assessment of key technical standards for grid connection of DERs, (iii) an extensive grid code literature review in both Europe and North America, and (iv) anticipated developments in grid requirements for DERs. The anticipated developments involve recent undertakings by prominent organizations, such as a recommended revision of IEEE Std. 1547 by the NERC and the upcoming release of Std. 1547.7 and 1547.8 by the IEEE. Based on this background work, the developed grid code outlines requirements governing key performance expectations of grid-connected systems. The requirements are allocated into three categories: (i) FRT capabilities, (ii) provision of ancillary services, and (iii) basic power quality constraints. These categories were identified as the anticipated core features of future DERs specific codes. The requirements pertaining to FRT and ancillary services are based primarily on Germany's stringent medium-voltage grid code but include several modifications to align with codes. The scope of DER systems addressed by the developed grid code include systems having multiple energy sources and local loads, with the requirements being enforced at a single PCC. Existing codes do not address such topologies as they generally focus on systems having either a single DER unit directly connected at the PCC, or multiple DER units that are identical in design and are usually connected in a simple parallel manner at the PCC. Furthermore, existing codes do not address DER systems having on-site loads that are of significant capacity in relation to the generating sources.

4.8 QUESTIONS

1. What is grid code?
2. What are the goals derived from the grid code?
3. What is the point of common coupling?
4. What is agreed active power?
5. What are ancillary services?
6. What is harmonic compensation?
7. What is black start capacity?
8. What is the meaning of voltage sag?
9. What is fault-ride-through and its requirement?

10. What is low-voltage ride-through?
11. What is high-voltage ride-through?
12. What is the IEEE standard 929 for PV systems?
13. What is maximum time shut down a PV plant when grid voltage goes abnormal?
14. What is the IEEE standard for interconnecting distributed resources with electric power systems?
15. What is the grid interconnection system response time when abnormal voltages are seen?
16. What is the Canadian electrical standard?
17. What is frequency operating limits for distributed resources?
18. What is current total demand distortion?
19. What is the current and voltage distortion limit?
20. What is the WES grid code?
21. What is DER-specific codes?
22. What is medium voltage grid code?
23. How active power is controlled in grid-connected PV systems?
24. How reactive power is controlled in grid-connected PV systems?
25. What is dynamic voltage support in grid-connected PV systems?

REFERENCES

[1] IEEE, "IEEE 1547,2008," in *IEEE Standard for Interconnecting Distributed Resources with Electric Power Systems*, 2008.

[2] IEEE, "IEEE Std 1547, 2014," in *IEEE Standard for Interconnecting Distributed Resources with Electric Power Systems*, 2014.

[3] IEEE, "IEEE STD 1547–2018," in *IEEE Standard for Interconnection and Interoperability of Distributed Energy Resources with Associated Electric Power Systems Interfaces*, 2018. doi: 10.1109/IEEESTD.2018.8332112.

[4] Ontario Power Authority (OPA), "Standard Definitions, Ontario Power Authority," 2010.

[5] Hydro One Networks, "Distributed Generation Technical Interconnection Requirements Interconnections at Voltages 50kv and Below," 2012.

[6] P. E. Sutherland, "Canadian grid codes and wind farm interconnections," in *2015 IEEE/IAS 51st Industrial & Commercial Power Systems Technical Conference (I&CPS)*, May 2015, pp. 1–7. doi: 10.1109/ICPS.2015.7266411.

[7] A. Foss and K. Leppik, "Design and implementation of an anti-Islanding protection strategy for distributed generation involving multiple passive protections," in *2009 IEEE Electrical Power & Energy Conference (EPEC)*, Oct. 2009, pp. 1–4. doi: 10.1109/EPEC.2009.5420963.

[8] T. Funabashi, Ed., *Integration of Distributed Energy Resources in Power Systems*. Elsevier, 2016. doi: 10.1016/C2014-0-03911-1.

[9] G. B. Gharehpetian and S. M. M. Agah, *Distributed Generation Systems*. Oxford: Elsevier, 2017. doi: 10.1016/C2014-0-03902-0.

[10] V. van Thong, J. Driesen, and R. Belmans, "Using distributed generation to support and provide ancillary services for the power system," in *2007 International Conference on Clean Electrical Power*, May 2007, pp. 159–163. doi: 10.1109/ICCEP.2007.384204.

[11] P. Strauss et al., "International white book on the grid integration of static converters," in *2009 10th International Conference on Electrical Power Quality and Utilisation*, Sep. 2009, pp. 1–6. doi: 10.1109/EPQU.2009.5318815.

[12] M. T. Aung and J. V. Milanovic, "The influence of transformer winding connections on the propagation of voltage sags," *IEEE Trans. Power Deliv.*, vol. 21, no. 1, pp. 262–269, Jan. 2006, doi: 10.1109/TPWRD.2005.855446.

[13] W. R. Mendes, M. I. Samesima, and F. A. Moura, "Influence of power transformer winding connections on the propagation of voltage sags through electric system," in *2008 5th International Conference on the European Electricity Market*, May 2008, pp. 1–6. doi: 10.1109/EEM.2008.4579012.

[14] C. Feltes, S. Engelhardt, J. Kretschmann, J. Fortmann, F. Koch, and I. Erlich, "High voltage ride-through of DFIG-based wind turbines," in *2008 IEEE Power and Energy Society General Meeting - Conversion and Delivery of Electrical Energy in the 21st Century,* Jul. 2008, pp. 1–8. doi: 10.1109/PES.2008.4596803.

[15] L. D. Ndoumbe, S. Eke, C. H. Kom, A. T. Yeremou, A. Nanfak, and G. M. Ngaleu, "Power quality problems, signature method for voltage dips and swells detection, classification and characterization," *Eur. J. Electr. Eng.*, vol. 23, no. 3, pp. 185–195, Jun. 2021, doi: 10.18280/ejee.230303.

[16] M. Prasad and A. Akella, "Comparison of voltage swell characteristics in power distribution system," *Int. J. Electr. Electron. Res.*, vol. 4, no. 3, pp. 67–73, Sep. 2016, doi: 10.37391/IJEER.040302.

[17] Institute of Electrical & Electronics Engineers, "IEEE 929–2000 Recommended Practice for Utility Interface of Photovoltaic (PV) Systems," 2000.

[18] T. S. Basso and R. DeBlasio, "IEEE 1547 series of standards: Interconnection issues," *IEEE Trans. Power Electron.*, vol. 19, no. 5, pp. 1159–1162, Sep. 2004, doi: 10.1109/TPEL.2004.834000.

[19] IEEE, "IEEE Std 1547, 2003," in *IEEE Standard for Interconnecting Distributed Resources with Electric Power Systems*, 2003. doi: 10.1109/IEEESTD.2003.94285.

[20] X. Tong, M. Zhong, X. Zhang, J. Deng, and Z. Zhang, "Voltage regulation strategy of AC distribution network based on distributed PV grid-connected inverter," *J. Eng.*, vol. 2019, no. 16, pp. 2525–2528, Mar. 2019, doi: 10.1049/joe.2018.8680.

[21] P. M. S. Carvalho, P. F. Correia, and L. A. F. Ferreira, "Distributed reactive power generation control for voltage rise mitigation in distribution networks," *IEEE Trans. Power Syst.*, vol. 23, no. 2, pp. 766–772, May 2008, doi: 10.1109/TPWRS.2008.919203.

[22] B. Saint, "Update on IEEE 1547 Series of Standards for distributed resources interconnection," in *PES T&D 2012*, May 2012, pp. 1–5. doi: 10.1109/TDC.2012.6281696.

[23] M. A. Khan, A. Haque, V. S. B. Kurukuru, and M. Saad, "Islanding detection techniques for grid-connected photovoltaic systems-A review," *Renew. Sustain. Energy Rev.*, vol. 154, p. 111854, Feb. 2022, doi: 10.1016/j.rser.2021.111854.

[24] Canadian Standards Association, "CSA C22.3 No. 9 Interconnection of Distributed Energy Resources and Electricity Supply Systems," Toronto, Canada, 2020.

[25] UL Standard, "UL-1741-Inverters, Converters, Controllers and Interconnection System Equipment for Use With Distributed Energy Resources," 2010.

[26] A. Taheri Kolli and N. Ghaffarzadeh, "A novel phaselet-based approach for islanding detection in inverter-based distributed generation systems," *Electr. Power Syst. Res.*, vol. 182, no. September 2019, 2020, doi: 10.1016/j.epsr.2020.106226.

[27] T. M. Blooming, N. Carolina, and D. J. Carnovale, "Application OF IEEE STD 519–1992 Harmonic Limits," pp. 1–9, 1992.

[28] V. S. B. Kurukuru, M. A. Khan, and R. Singh, "Performance optimization of UPFC assisted hybrid power system," *2018 IEEMA Eng. Infin. Conf. eTechNxT 2018*, pp. 1–6, 2018, doi: 10.1109/ETECHNXT.2018.8385295.

[29] M. Tsili and S. Papathanassiou, "A review of grid code technical requirements for wind farms," *IET Renew. Power Gener.*, vol. 3, no. 3, p. 308, 2009, doi: 10.1049/iet-rpg.2008.0070.

[30] I. M. de Alegría, J. Andreu, J. L. Martín, P. Ibañez, J. L. Villate, and H. Camblong, "Connection requirements for wind farms: A survey on technical requierements and regulation," *Renew. Sustain. Energy Rev.*, vol. 11, no. 8, pp. 1858–1872, Oct. 2007, doi: 10.1016/j.rser.2006.01.008.

[31] Hydro-Québec Transmission System, "Technical Requirements for the Connection of Generating Stations to the Hydro-Québec Transmission System," 2018.

[32] J. H. Kehler, D. McCracnk, and P.Eng, "Wind Power Facility Technical Requirements," 2004. [Online]. Available: Wind-Power-Facility-Technical-Requirements-Revision0-signatures-JRF.pdf

[33] S. Bernard, D. Beaulieu, and G. Trudel, "Hydro-Quebec grid code for wind farm interconnection," in *IEEE Power Engineering Society General Meeting, 2005*, pp. 2786–2790. doi: 10.1109/PES.2005.1489702.

[34] S. Perlenfein, M. Ropp, J. Neely, S. Gonzalez, and L. Rashkin, "Subharmonic power line carrier (PLC) based island detection," in *Conference Proceedings – IEEE Applied Power Electronics Conference and Expoosition – APEC*, vol. 2015, no. May, pp. 2230–2236, 2015, doi: 10.1109/APEC.2015.7104659.

[35] M. A. Khan, A. Haque, and V. S. Bharath, "Dynamic voltage support for low voltage ride through operation in single-phase grid-connected photovoltaic systems," *IEEE Trans. Power Electron.*, vol. 8993, no. c, 2021, doi: 10.1109/TPEL.2021.3073589.

[36] M. A. Khan, A. Haque, and V. S. B. Kurukuru, "Droop based Low voltage ride through implementation for grid integrated photovoltaic system," in *2019 International Conference on Power Electronics, Control and Automation (ICPECA)*, Nov. 2019, pp. 1–5. doi: 10.1109/ICPECA47973.2019.8975467.

[37] M. A. Khan, V. S. B. Kurukuru, and A. Haque, "Islanding classification and low-voltage ride through for grid connected transformerless inverter," in *Intelligent Circuits and Systems*, 1st Edition. Florida: CRC Press, 2021, p. 7.

[38] A. Notholt, "Germany's new code for generation plants connected to medium-voltage networks and its repercussion on inverter control," in *International Conference on Renewable Energies and Power Quality*, 2009, vol. 1, no. 7, pp. 716–720. doi: 10.24084/repqj07.482.

[39] H. Berndt, M. Hermann, H. Kreye, R. Reinisch, U. Scherer, and J. Vanzetta, "TransmissionCode 2007 Network and System Rules of the German Transmission System Operators," Berlin: 2007.

[40] T. Degner, G. Arnold, and M. Braun, "Utility-scale PV systems: Grid connection requirements, test procedures and European harmonization," 2009. [Online]. Available: http://www.der-lab.net/downloads/pvi4-08_3.pdf

[41] P. M. de Almeida, P. G. Barbosa, C. A. Duque, and P. F. Ribeiro, "Grid connection considerations for the integration of PV and wind sources," in *2014 16th International Conference on Harmonics and Quality of Power (ICHQP)*, May 2014, pp. 6–9. doi: 10.1109/ICHQP.2014.6842908.

[42] H. Leite, P. Ramalho, B. Silva, and R. Fiteiro, "Distributed generation protection scheme to permit 'ride-through fault'," in *CIRED 2009–20th International Conference and Exhibition on Electricity Distribution – Part 1*, Jun. 2009, pp. 1–4.

[43] E. Troester, "New German grid codes for connecting PV systems to the medium voltage power grid," in *2nd International Workshop on Concentrating Photovoltaic Power Plants: Optical Design, Production, Grid Connection*, 2009, pp. 9–10.

[44] California Solar Initiative (CSI). https://www.cpuc.ca.gov/industries-and-topics/electrical-energy/demand-side-management/california-solar-initiative (accessed Jul. 18, 2022).

[45] Pacific Gas and Electric Company, "Electric Rule no. 21 Generating Facility Interconnections," 2021.

[46] M. Milligan, "Sources of grid reliability services," *Electr. J.*, vol. 31, no. 9, pp. 1–7, Nov. 2018, doi: 10.1016/j.tej.2018.10.002.

[47] J. Forrester, "The value of CSP with thermal energy storage in providing grid stability," *Energy Procedia*, vol. 49, pp. 1632–1641, 2014, doi: 10.1016/j.egypro.2014.03.172.

[48] D. L. Brooks and M. Patel, "Panel: Standards; interconnection requirements for wind and solar generation NERC integrating variable generation task force," in *2011 IEEE Power and Energy Society General Meeting*, Jul. 2011, pp. 1–3. doi: 10.1109/ PES.2011.6039360.

[49] BDEW, "Generating Plants Connected to the Medium-Voltage Network," *Tech. Guidel.*, Berlin, Germany, no. June, 2008.

[50] N. Mohan, T. M. Undeland, and W. P. Robbins, *Power Electronics: Converters, Applications, and Design.* John Wiley & Sons, Inc., 2002.

[51] A. Anzalchi and A. Sarwat, "Overview of technical specifications for grid-connected photovoltaic systems," *Energy Convers. Manag.*, vol. 152, no. August, pp. 312–327, 2017, doi: 10.1016/j.enconman.2017.09.049.

[52] Canadian Standards Association, "CSA C235 Preferred Voltage Levels for AC Systems up to 50000 V," 2019.

[53] "Technical Guideline for Interconnection of Generators to the Distribution System – EEP." https://electrical-engineering-portal.com/download-center/books-and-guides/ power-substations/interconnection-generators (accessed Jul. 18, 2022)

5 Standalone Control Operation of PV Inverter

5.1 INTRODUCTION

With the increase in demand for power all over the world and the constant depiction of non-renewable resources, there is a need for a solution for sustainable power generation. With the emergence of solar energy as one of the best options in recent years, many advancements have been taken for the utilization of most of the energy generated. A transformerless inverter is one of the improvements made in the field of inverters. The main aim of the transformerless inverter is to eliminate the losses taking place in the inverter due to the presence of the transformer, and as an outcome, it also leads to a decrease in the size of the inverter system. The only issue with the transformerless inverter stands in the form of leakage current due to the absence of isolation between the AC and DC halves of the inverters.

A study by Xiao et al. [1] proposes an optimized structured full-bridge inverter along with two extra switches and a voltage divider capacitance which tends to assure a freewheeling path for the clamped input. The research by Patrao et al. [2] presents an overview of grid-connected transformerless inverters topologies, and a comparison is carried out with the conventional topologies. Gubia et al. [3] studied the common-mode-related issues in single-phase transformerless inverters. The presence of ground current and its effect on efficiency and grid current quality is discussed. Rodriguez et al. [4] proposed a need for transformerless inverters with step-up output. The proposed topology with seven layers claims to have obtained higher efficiency even by decreasing the number of switches in the application. A study by Freddy et al. [5] proposed a modified H-bridge rectifier operating in a zero state, and the combination led to lower losses in AC decoupling and also eradicated leakage current. Azri et al. [6] proposed a bipolar sinusoidal pulse width modulation (SPWM) for a transformerless inverter with the aim of reducing the leakage current and its effectiveness was compared with the unipolar SPWM. A study done by Vazquez et al. [7] aims to provide a link between the electric grid and PV generator with the reduced common-mode current. Even conduction losses and switching losses are taken into consideration while designing the modulation strategy. Hu et al. [8] proposed a transformerless inverter modified from highly efficient and reliable inverter concept (HERIC) topology consisting of a clamp cell so that it can eliminate the leakage current. Islam et al. [9] focused on studying various transformerless topologies, and their classification was carried out for them by leakage current, a rating of the component, advantages, and disadvantages. Fu et al. [10] proposed a vector control method based on neural networks for inverters with LCL filters. Based on adaptive dynamic programming, the neural network is trained and optimal control is carried out.

DOI: 10.1201/9781003257189-5

In 2016, Dutta et al.'s [11] work focused on the performance-based analysis of full-bridge grid-connected transformerless inverters depending upon the efficiency, power loss, and THD for different conditions. Pranav et al. [12] proposed using a fuzzy controller for increasing the power of the photovoltaic system, and the current is controlled in the circuit using a proportional-resonant controller. Fu et al. [13] studied a comparative analysis of grid-connected inverters operating with L, LC, and LCL filters. Neural network-based vector control is also implemented in the system. Li et al. [14] presented an H5 D topology of transformerless inverters and worked on obtaining the complete elimination of common-mode current from the system. Khan et al. [15] proposed a diode-free common-mode voltage clamping and freewheeling branch for a transformerless inverter for eliminating leakage current and reducing conductance loss. Somani et al.'s [16] work proposed a grid-connected inverter with controlled power and a high total harmonic distortion (THD) value as a result of the HERIC topology. Sun et al.'s [17] study focused on a transformerless inverter topology with the implementation of a new freewheeling loop that compensates for the galvanic isolation of conventional inverters and protects the system from leakage current. Xiao et al. [18] proposed a zero-voltage transition of HERIC topology in which, by integrating freewheeling switches and a resonant tank, a small network of resonance is formed. Ahmad et al. [19] reviewed different transformerless inverter topologies, and a performance comparison of different topologies were performed by common-mode voltage, THD, leakage current, efficiency, and device losses.

Controlling a particular parameter of an inverter, Maswood et al. [20] researched a proportional and integral controller implemented in a voltage source inverter for controlling pulse width modulation. Selvaraj et al. [21] proposed the implementation of a PI controller for grid-connected PV systems. A PI controller is used for providing robust control with an operating point near the unity power factor. Ayob et al. [22] submitted a brief report on the application of fuzzy training value of PI gains for controlling the inverter. Villanueva et al. [23] presented an adaptive control scheme for providing maximum power-point tracking, independent of each DC link. It offers advantages such as low switching frequency and current ripple in comparison with the conventional system. The work by Shen et al. [24] provided insight into proportional and resonant controllers for current control in the case of a grid-connected inverter using an LCL filter. Khateb et al. [25] discussed proportional, integral, and derivative (PID) for controlling the single-phase inverter along with the single-ended DC–DC converter. Ziegler-Nichols method is utilized for optimization purposes. A study done by Kerekes et al. [26] proposed a new topology with a diode clamped at the midpoint of an H-bridge. Performance analysis was done by comparing it with existing topologies. Buticchi et al. [27] proposed a digital control along the PWM of a transformerless PV system with the aim of reducing the ground leakage current and compensating for switching losses. Yang et al. [28] studied low-voltage ride-through under grid fault and even presented a reactive power injection as a control strategy. Hassaine et al. [29] provided an overview of power inverter topologies and their control structure implementation. The system investigates the various control structure. Work done by Wai et al. [30] proposed a study of two different control strategies – total slide mode controller and adaptive fuzzy neural network control for voltage tracking to be considered in a single stage of boost inverter and finding experimental results for the same. Sefa et al. [31] have implemented a fuzzy-PI controller for interactive grid systems. Fuzzy controllers are used for tuning the PI controller, and experimental testing is carried out. Ahuja et al. [32] performed power controlling of a grid-connected inverter using an adaptive-network-based fuzzy inference system for enhancing the performance quality and dynamics of the system. Rajeev et al. [33] proposed a closed-loop system for a novel transformerless inverter topology for controlling the grid power and reactive power by taking maximum power and grid synchronization into consideration. Mohan et al. [34] designed a multi-level inverter, where controlling different levels of the inverter is handled by the neuro-fuzzy controller.

The review presented by Chakraborty et al. [35] offered an overview of different transformerless inverter topologies along with consideration for various filter topologies considered for load applications. The comparison is carried out by size, efficiency, cost, and losses. Another publication by Islam et al. [36] presented a new transformerless inverter topology along with super-junction MOSFETs and SiC diodes, and an attempt was made to operate switches at unity power. A reduction of leakage current is obtained in the process. Fei et al. [37] presented two-stage single-phase grid-connected solar inverter control taking place by the implementation of an adaptive fuzzy sliding-mode controller. Trinh et al. [38] proposed a current control approach by the application of a three-vector PI controller system. Anantwar et al. [39] presented a self-turned PI controller for the stabilization of the system's active and reactive power.

The research presented here focuses on the stability analysis and control of transformerless inverters, and a comparison is done on various performance parameters. A novel topology is proposed which is compared with the existing topologies regarding performance parameters. A brief description of different transformerless inverter topologies is presented in Section 5.2, whereas the control algorithm implemented is explained in Section 5.3. Section 5.4 presents simulation and discussion related to experimental implementation, and in Section 5.5, results are presented in brief.

5.2 TRANSFORMERLESS INVERTER TOPOLOGIES

Recently, there has been sustainable development in the field of power electronics converters with an aim of reducing losses and making them more efficient. In the section below, few of transformerless inverter topologies are explained.

5.2.1 H5 TOPOLOGY

The H5 topology is depicted in Figure 5.1, which consists of five active switches. A 50 switch is present in between the positive half of the DC bus and the H4 topology. The fifth switch operates at a higher or grid frequency. S1 remains on during positive grid voltage, and a unipolar voltage grid is generated by switching between S5 and S3. For negative voltage, S1 is turned on, S4 and S5 commute at higher frequencies, and S2 and S4 commute complementary to each other, forming a freewheeling path. The commutation of S1 takes place complementary to S3 which tends to create a freewheeling path. The commutation of the S5 switch takes place in such a way that

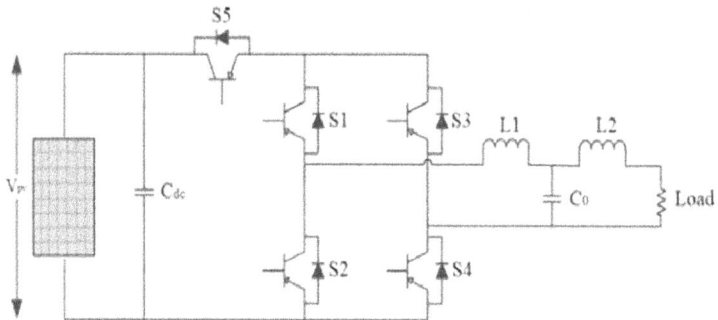

FIGURE 5.1   Representation of H5 topology.

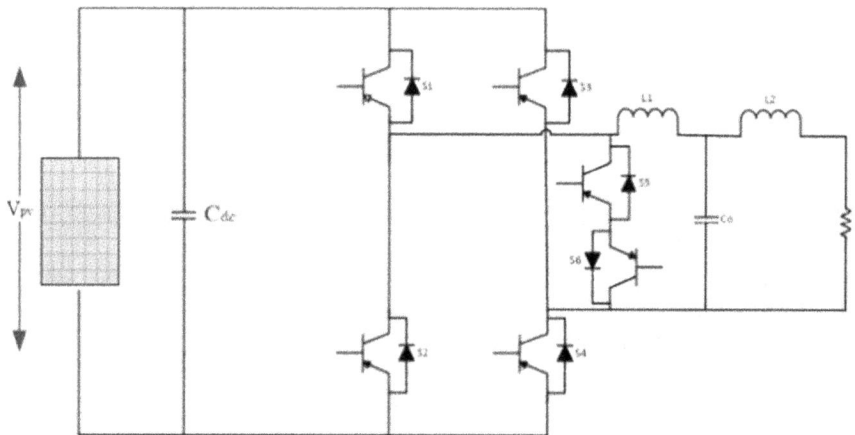

FIGURE 5.2   HERIC topology representation.

during a zero-voltage state, it is turned off and the PV panel is isolated from the grid and AC load. It leads to a reduction in the leakage current. The commutation of S5 takes place at double the frequency, which results in losses of switching and an unbalanced condition. This problem is rectified by designing the heat sink, and the power density is affected. In the H5 topology, the leakage current is low, but it is not eliminated entirely.

5.2.2 HERIC TOPOLOGY

The HERIC topology as depicted in Figure 5.2 consists of six switches. Two switches are connected back to back before the filtering of the H4 topology of the inverter. Depending upon the polarity of the grid, voltage S5 and S6 switches of the inverters are turned on and off at the zero-voltage state. All four switches S1, S2, S3, and S4 are turned off during the zero-voltage state, isolating the PV module from the AC load and the grid. For the positive half of the operation, S6 is switched on, whereas S1 and S4 are switched on simultaneously at a switching frequency, and S2 and S3 remain off. When the active vector is present, S1 and S4 are on, and the load is connected to the grid, whereas, in the zero-state vector, the current flow via S6 and diode at the S5 switch and keeping S1 and S4 disconnected. This causes a freewheeling period. For the negative half of the circuit, S2 and S3 are turned on by connecting the load from the PV module. For the zero-voltage vector part, current flow from S5 and the diode of S6 caused the negative half of the freewheeling circuit. The leakage current is small but not eliminated completely.

5.2.3 PROPOSED TOPOLOGY

The proposed topology is depicted in Figure 5.3. In the proposed topology, a switch S5 is added to the DC link part of the inverter and an anti-parallel switch is added to the filter end of the inverter. S5, S6, and S7 operate on the grid frequency, and the polarity of the switch determines the system's polarity. Different levels of voltage are created by the anti-parallel switches, providing isolation between the DC and AC halves of the system. As a result, the leakage current between the source and load end is decreased rapidly. S1 and S3 switches operate at the switching frequency alternatingly depending on the poles of the system. S1 remains on during positive grid voltage, and a unipolar voltage grid is generated by switching between S5 and S3.

For the positive half of the operation, S6 is switched on, whereas S1 and S4 are turned simultaneously at a switching frequency, and S2 and S3 are kept off, as shown in Figure 5.4a. When the active vector is present, S1 and S4 are on and the load is connected to the grid, whereas, in the zero-state vector, the current flow via S6 and the diode at the S7 switch and keeping S1 and S4 off disconnected causes a freewheeling period, as shown in Figure 5.4b.

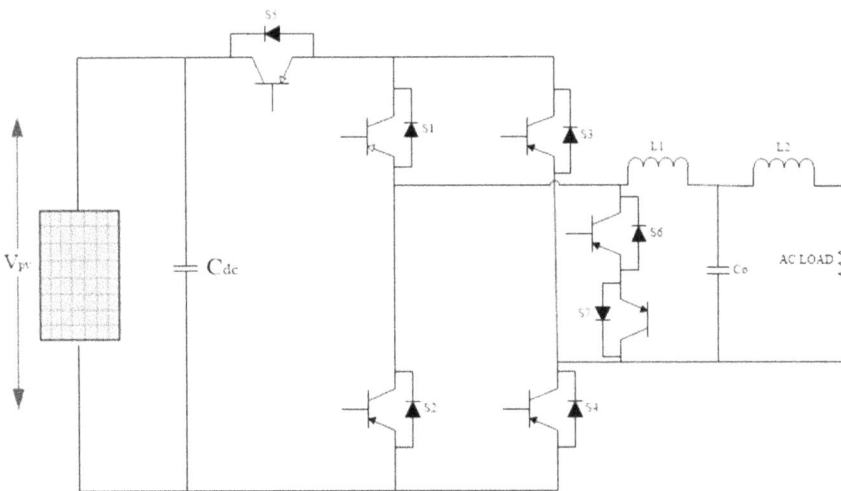

FIGURE 5.3   Proposed topology representation.

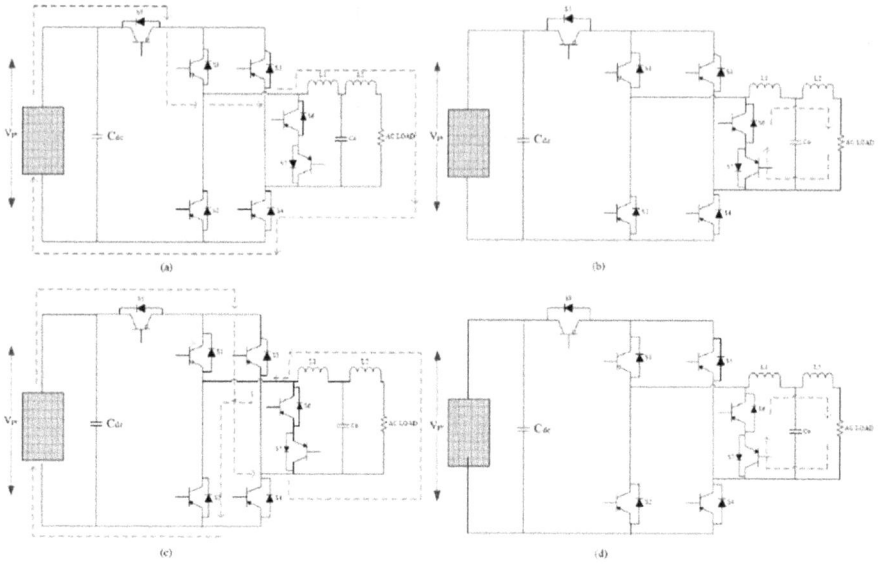

FIGURE 5.4   Different modes of the proposed topology.

For negative voltage, S1 is turned on, S4 and S5 commute at a higher frequency, and S2 and S4 commute complementary to each other forming a freewheeling path. The commutation of S1 takes place complementary to S3 which tends to create a freewheeling path. The compensation of the S5 switch takes place in such a way that during a zero-voltage state, it is turned off and the PV panel is isolated from the grid and AC load. For the negative half of the circuit, S2 and S3 are turned on, connecting the load from the PV module as shown in Figure 5.4c. For the zero-voltage vector part, the current flow from S7 and the diode of S6 causes the negative half of the freewheeling circuit, as shown in Figure 5.4d.

5.3   CONTROL ALGORITHMS

The implementation of feedback control is carried out by using PID, fuzzy, adaptive neural, and fuzzy inference systems (ANFIS), fuzzy PID, and ANFIS-PID controllers in the system. A load connection is provided at the output end of the inverter [40–42]. The output of the system is detected with the help of a voltage sensor, and the reference signal is compared with it for obtaining the error of the system. Depending on the different controllers, the error is then processed, and failure is minimized for the system. The error signal is then compared with the triangular carrier signal, and intersecting helps in obtaining the pulse width for switching. Different controllers applied in the system are explained in the sections below. Control implementation for the system is illustrated in the block diagram in Figure 5.5.

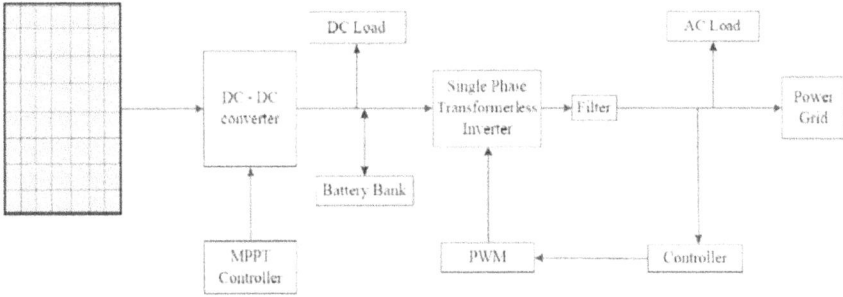

FIGURE 5.5 Block diagram representation of control implementation of the inverter.

5.3.1 PID CONTROLLER

A PID controller is one of the most convenient modes of control. Previously, the implementation of the PI controller has taken place on the DC–DC half of the converter system [25], and in some research studies, digital PI control was explained for the control of the grid-connected inverter [21]. However, there is not any detailed analysis focusing on the settling time after controller or voltage error calculation, and even the transformerless inverter has not been considered. The different gain factors of the proportional (K_p), integral (K_i), and differentiator (K_d) are varied for the system to obtain a stable voltage output for varying load and ambient conditions.

The general transfer function of PID control is

$$\text{T.F.} = K_p\left[1 + \frac{1}{T_i s} + T_d s\right] \tag{5.1}$$

where K_p=proportional gain, T_i=time constant of the integral, and T_d=time constant of the derivative. For the control output, the following equation is considered.

$$U_n = r_p e + r_i \int_0^t e\,dt + r_d e' \tag{5.2}$$

where $r_p = K_p$, $r_i = \dfrac{K_p}{T_i s}$, and $r_d = K_p T_d$

$$e = U_{ref} - U_{out} \tag{5.3}$$

where U_{ref}=desired gain value and U_{out}=output gain value.

The error of the system is observed to be reduced with an increase in the proportional gain, whereas the integrator gain minimizes the offset value of the controller. Proportional gain is determined by the amount of present gain, whereas the past error value determines the gain of the integrator. Tuning of the PID controller is required for obtaining a better response of the system. The PID controllers are fed with the system error which was presented by comparing the reference and the actual signal. As given by the amplitude of the triangular waveform, the error need not be more than the range.

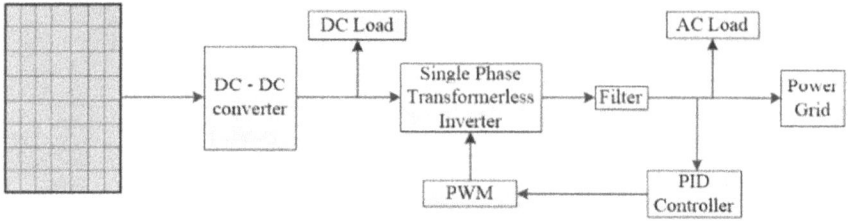

FIGURE 5.6   Block diagram representation of PID implementation of the inverter.

In the inverter topologies discussed in the sections before, PID control was implemented in the error of the system and a pulse was generated by feeding the PID control output to the pulse width modulator so that the feedback PID control signal can control the switching of the different inverter switches. The implementation is further illustrated in Figure 5.6.

5.3.2 Fuzzy Logic Controller

The aim of the control techniques was to focus on the concept that human brain thinking has no limit. Not a lot of research has been carried out focusing on different performance parameters. Linguistic terms which are not defined precisely are known as fuzzy sets. It may not always be groups or classes of objects for classification. Several groups can contain various objects at the same time. The criteria of a group define various purposes. Precision should be maintained in different class criteria. Classic logic may not be able to determine some of the objects.

Fuzzy sets are classified by three processes as explained below:

During the fuzzification process, the crisp value input is converted into linguistic terms, which in the later part are made qualified for the membership function of the fuzzy system using the fuzzy set. The variable can be in the form of a word or statement instead of a numerical value. During fuzzification, the real scalar value is converted into a fuzzy value, and this leads to the conversion of the crisp value in the form of a group or linguistic variable. Once the fuzzification is completed, it is passed through a specific set of rules to determine the output for a value of the input. Rules can be in the relationship between multi-inputs to establish an output. Once the output is obtained, the defuzzification process concludes the fuzzy operation by converting the fuzzy value into a crisp value.

Research carried out here focuses on the application of fuzzy on different transformerless inverter topologies so that the output voltage of the system can be made stable even when the load and ambient conditions are varied. For control implementation, a block structure is followed as shown in the figure. The fuzzy structure and the membership function are illustrated in Figure 5.7.

The rules set with fuzzy membership function are shown in Figure 5.8.

In the inverter systems, the error from reference and the actual value is fed to the fuzzy toolbox where different membership functions are defined for error and change in error, and depending on the relationship, a set of rules is determined as discussed

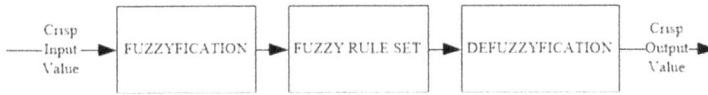

FIGURE 5.7 Fuzzy analysis structural breakdown.

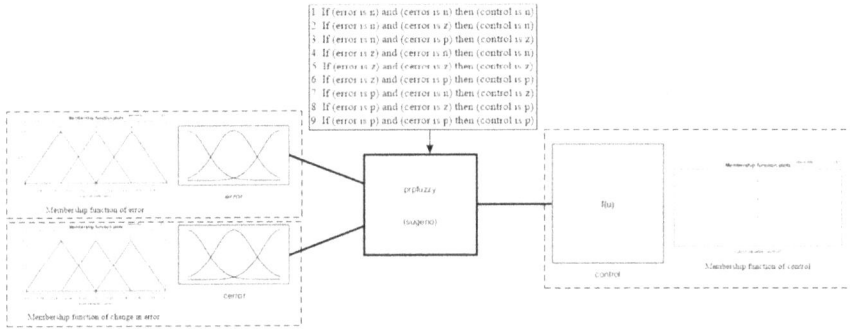

FIGURE 5.8 Fuzzy membership function and rule execution of the inverter.

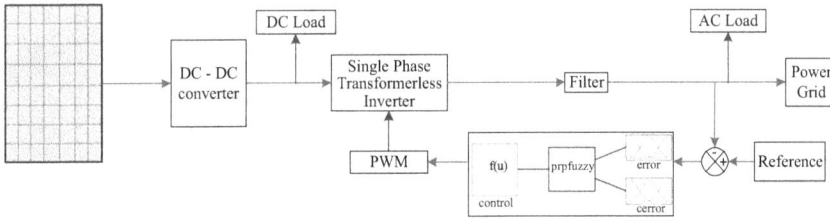

FIGURE 5.9 Block diagram representation of a fuzzy implementation of the inverter.

above. When the crisp error value is converted into a linguistic variable, and the same is done for the change in the error value, they are compared for the output as per the fuzzy rules. The output of the fuzzy toolbox is then fed to the pulse width modulator, and depending on the pulse, the switching of the different power electronic switches is controlled. The implementation is further illustrated in Figure 5.9.

5.3.3 ANFIS CONTROLLER

The adaptive neural fuzzy inference system is commonly known as ANFIS. The ANFIS is a relatively more advanced controller than fuzzy and PID, not much research related to inverter control has been done before. The ANFIS along with the proposed topology tends to give better results than most of the existing work. Depending on the data set of input and output, the ANFIS toolbox function is constructed on the fuzzy inference system whose tuning of the membership function is carried out by the back-propagation algorithm or a combination of the method of an at least square type is presented. The above-modeled data learns from the fuzzy system.

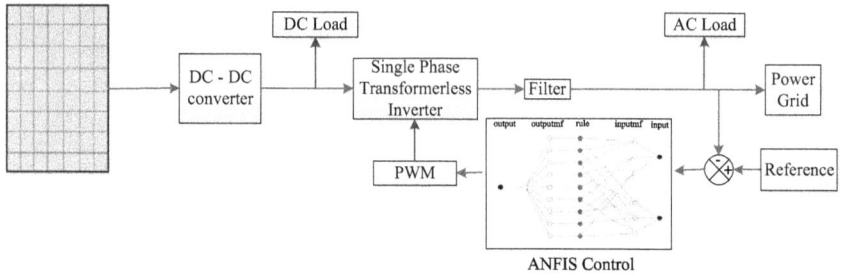

FIGURE 5.10   Block diagram representation of ANFIS implementation of the inverter.

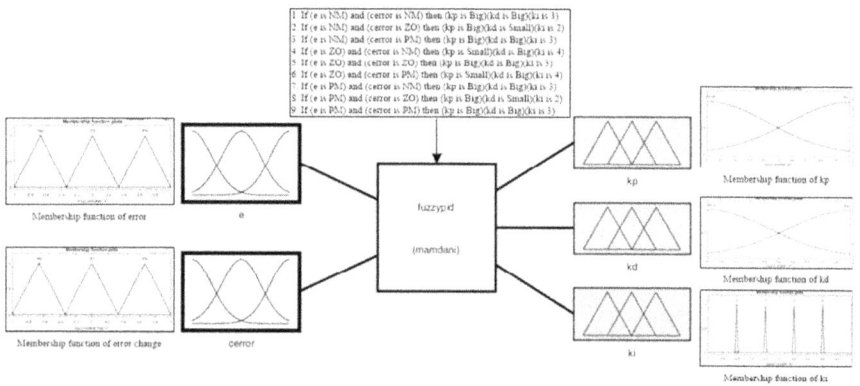

FIGURE 5.11   Fuzzy membership function and rule execution of the inverter.

In the topologies presented in the research, the ANFIS is implemented with the error of the system which trains the value of membership function and rules set presented as in the fuzzy system. The output of the ANFIS is then fed into the pulse width modulator, and the generated output pulse then controls the switching operation of the power electronic switches of the inverter. In Figure 5.10, the working of the feedback control is illustrated.

5.3.4 Fuzzy PID Controller

Fuzzy PID is a self-tuning method in which fuzzy rules are used for tuning the PID controller. Previously, fuzzy PID was used by few researchers for controlling the inverter [31,39], but very few have focused on transformerless inverter topology working along with the fuzzy PID controller and its performance analysis. It helps in obtaining system stability with more reliability, accuracy, and better performance. The necessary approach aims at finding the error in the system and utilizing it for obtaining different gain values for the PID controller. In fuzzy controllers, fuzzification of error and change in error is carried out as per the rules, and defuzzification of the output is performed. The fuzzy part of the control is depicted in Figure 5.11.

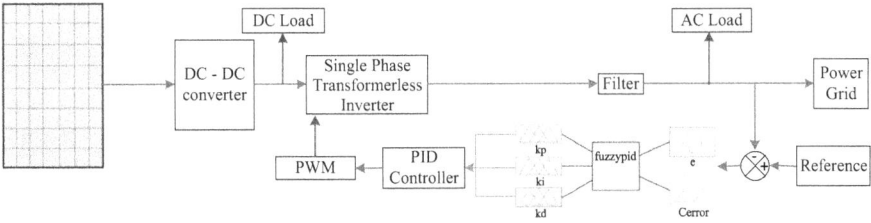

FIGURE 5.12   Block diagram representation of a fuzzy PID implementation of the inverter.

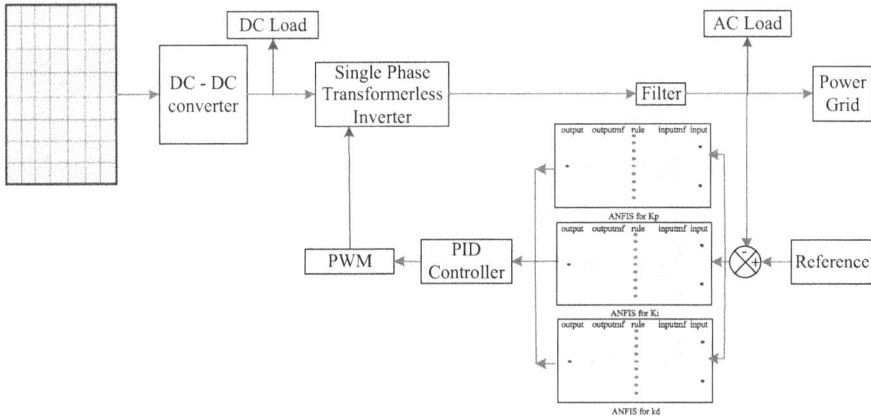

FIGURE 5.13   Block diagram representation of an ANFIS-PID implementation of the inverter.

The defuzzied output provided the Kp, ki, and kd for the PID controller and tuned the PID controller. The signal output of the PID controller is forwarded to the pulse width modulator of the system which compares it with triangular waves for obtaining switching signals for different power electronic switches present in the inverter. The control of the system with the fuzzy PID controller is illustrated in Figure 5.12.

5.3.5   ANFIS-PID CONTROLLER

For improving controller reliability for minimizing the error, fuzzy membership functions need to be defined, whereas ANFISs are implemented in parallel structures and embedded into control systems. The ANFIS-PID controller has not been explored a lot in terms of voltage regulation of inverters. In the following implementation, the ANFIS is computed for each of the gain components by using the error and finding the value of Kp, ki, and kd by implementing a different set of rules for various ANFIS applications. In Figure 5.13, the use of the ANFIS-PID in the inverter is depicted in the feedback application.

5.4 SIMULATION AND DISCUSSION

The simulation for studying different topologies and their response to various control methods was carried out on MATLAB 2015b. A separate component such as a solar panel, filter, and feedback control half of the system was modeled. In the sections below, different component designing types are explained.

5.4.1 SOLAR PANEL DESIGNING

A solar panel of 2 kW Sanyo (HIP-210HKHA6) was considered for design. Different input components such as irradiance and temperature are varied by keeping the maximum range up to 1000 W/m^2 and 25°C, respectively. Other parameters needed for designing are mentioned in Table 5.1.

5.4.2 FILTER DESIGNING

The filter is implemented on the output half of the inverter to remove the harmonics present in the output signal. A T-LCL filter was employed in the system which consists of two inductors, and one capacitor and is arranged in the form of a T shape. For calculating the value, the following formulas are implemented.

$$c = \frac{1}{2XpXfXZ} \tag{5.4}$$

$$L = CZ^2 \tag{5.5}$$

where
  f = cut-off frequency
  Z = impedance

5.4.3 FEEDBACK STABILITY ANALYSIS

The transfer function of the system is studied for stability analysis of the system. Analysis of different components of the system, such as the filter, inverter, pulse width modulator, and control unit, are analyzed, and the formation of the control block was done. A block representation of the system control unit is presented in Figure 5.14.

TABLE 5.1
Solar Panel Parameters

Specification	Value
Short-circuit current (I_{Short})	5.57 A
Maximum power (P_{-max})	210 W
Open-circuit voltage (V_{open})	50.0 V

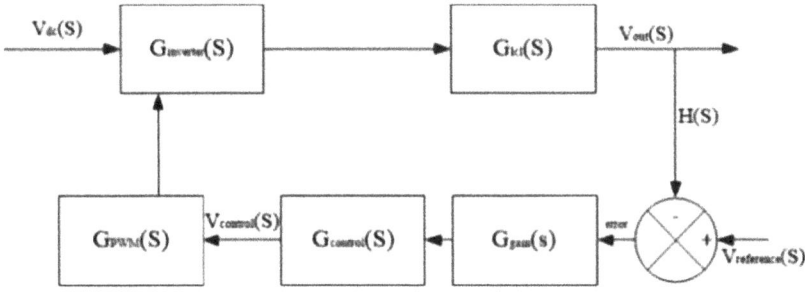

FIGURE 5.14 Control loop of the system.

The transfer function for determining the stability is as follows:

$$L.T = \frac{G_{\text{inverter}} G_{\text{lcl}}}{1 + (G_{\text{inverter}} G \text{ lcl}) G_{\text{control}} G_{\text{gain}} G_{\text{PWM}} H(s)} \qquad (5.6)$$

where

$G_{\text{inverter}} = $ gain of the inverter

$G_{\text{lcl}} = $ gain of filter $= \dfrac{1}{s(L_1 + L_2) + s^3(C_f L_1 L_2)}$

$G_{\text{control}} = $ gain of the PI controller $= K_p + \dfrac{K_p}{sT}$

$G_{\text{gain}} = $ overall gain

$G_{\text{PWM}} = $ Pulse width modulation gain $= \dfrac{1}{1 + 1.5Ts}$

As per the design of the inverter, it is expected that a 230 V output is to be obtained for the inverter even when there is a variation of load or changing ambient conditions. An acceptable range of 6% of error in voltage is considered for the inverter as recommended in the National Electrical Code, which is published by the Bureau of Indian standards [43–45]. Even the runtime load variation for different load profiles is reviewed to make sure that the stability time of the system is smaller and the accuracy in the voltage range is maintained. A study regarding THD was also done, and the THD acceptable range was to be kept at 5% as per EN 50160 [46]. Even the common-mode voltage analysis of the system was carried out on different topologies. As the control system is implemented for voltage stability, control system analysis of the system was performed using the Bode plot. In a separate section below, the results are explained in different categories which have been mentioned before.

5.4.4   INITIAL LOAD VARIATION (T = 0)

The primary concern of the work done in this paper is to maintain a stable voltage even when there is a variation in the load by the implementation of different control algorithms. In Figure 5.15, it was observed that a constant voltage was obtained around a 230 V reference value with some acceptable range of errors as discussed above. Even the initial settling time was mentioned in the figure for finding the time needed to get a system stable when it is turned on at a certain load.

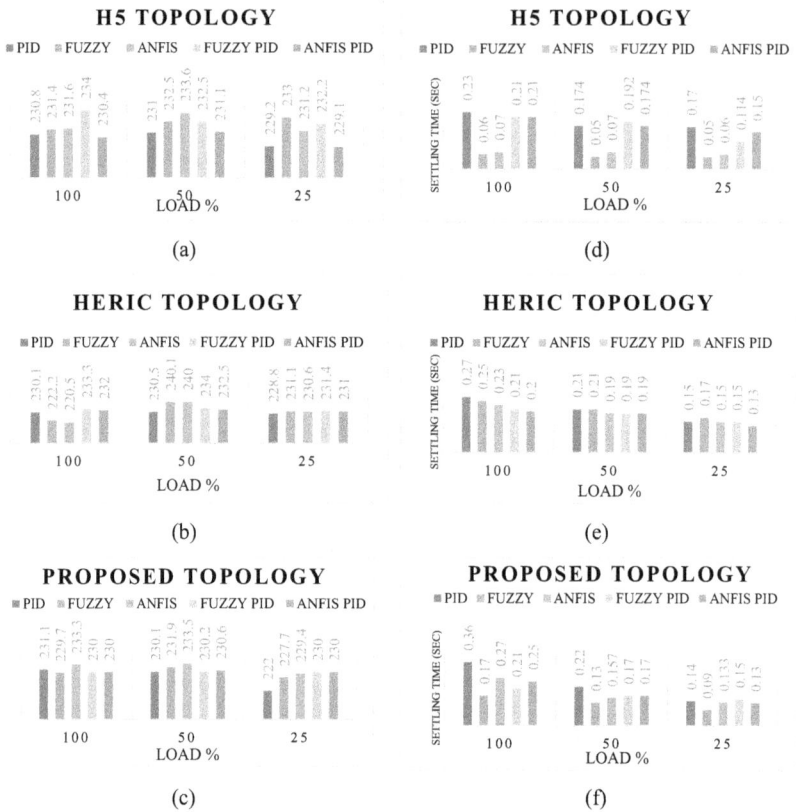

FIGURE 5.15 (a) Voltage at different loads for H5. (b) Voltage at different loads for HERIC. (c) Voltage at different loads for the proposed topology. (d) Settling time at the initial loading for H5. (e) Settling time at the initial loading for HERIC. (f) Settling time at the initial loading for the proposed topology.

As per the response time, it can be analyzed that in the H5 topology, the fuzzy and ANFIS controller gave a relatively fast settled output than that of other applied control techniques. In the case of HERIC topology, the settling times for all the different controllers are very close to each other, making ANFIS-PID the fastest in achieving stability. In the proposed topology, the settling time for the fuzzy controller is found to be the fastest, but even other controllers do not lag far behind.

In Tables 5.2–5.4, a different value of the error is calculated by keeping 230 V as a reference voltage. In Table 5.2, the H5 topology displays a very low error in the case of the ANFIS-PID controller, whereas the fuzzy PID has a higher error value which is still very much within the range of permitted error levels.

In the case of HERIC topology, there is a very high error when controlled using Fuzzy and ANFIC algorithms under 50% and 25% of the load as revealed in the table, but in the case of 100% load, the error is less. Superlative results were obtained for the PID controller for the corresponding topology, as can be seen in Table 5.3.

TABLE 5.2
H5 Topology Error Calculation

Load %	H5				
	PID	Fuzzy	ANFIS	Fuzzy PID	ANFIS-PID
100	0.3%	0.6%	0.6%	1.7%	0.1%
50	0.4%	1.0%	1.5%	1%	0.4%
25	0.3%	1.3%	0.5%	1.1%	0.3%

TABLE 5.3
HERIC Topology Error Calculation

Load %	HERIC				
	PID	Fuzzy	ANFIS	Fuzzy PID	ANFIS-PID
100	0.04%	3.3%	4.13%	1.434%	0.86%
50	0.21%	4.34%	4.34%	1.739%	1.08%
25	0.52%	0.47%	0.60%	0.608%	0.43%

TABLE 5.4
Proposed Topology Error Calculation

Load %	Proposed				
	PID	Fuzzy	ANFIS	Fuzzy PID	ANFIS-PID
100	0.47%	0.13%	1.43%	0%	0%
50	0.04%	0.82%	1.52%	0%	0.26%
25	3.47%	1%	0.26%	0%	0%

In the proposed topology illustrated in Table 5.4, fuzzy PID and ANFIS-PID give decent outcomes as the error is zero for the two-load profile, and only in case of the ANFIS algorithm, there is a rise in error, but even that is very insignificant in compared to the permitted range.

5.4.5 Runtime Load Variation (T = 0+)

As the load keeps on varying when the system is in operation, it is necessary for the system to retain its stability and operate at a constant voltage. The settling time plays a vital role, as during the operating condition, if the settling time is larger, then there may be a fault detection on the relay end which may even end up shutting down the system and causing a loss of money and a vast extensive blackout as well. Similarly, as the above analysis was done to obtain a stable output voltage, its settling time was studied for different topologies by the implementation of different control algorithms. In Table 5.5, it is observed that the settling time keeps on reducing as the load

TABLE 5.5

Settling Time During Operating Conditions (in a sec)

Load%	H5 Topology				HERIC Topology				Proposed Topology			
Control	0 to 25%	25 to 50%	50 to 75%	75 to 100%	0 to 25%	25 to 50%	50 to 75%	75 to 100%	0 to 25%	25 to 50%	50 to 75%	75 to 100%
PID	0.17	0.034	0.024	0.001	0.15	0.034	0.02	0.001	0.14	0.07	0.052	0.013
Fuzzy	0.05	0.037	0.024	0.02	0.17	0.05	0.03	0.005	0.09	0.05	0.009	0.004
ANFIS	0.06	0.04	0.037	0.032	0.15	0.07	0.034	0.001	0.133	0.06	0.052	0.006
Fuzzy PID	0.114	0.037	0.02	0.001	0.15	0.036	0.02	0.003	0.15	0.055	0.034	0.001
ANFIS-PID	0.15	0.03	0.025	0.004	0.13	0.037	0.017	0.006	0.13	0.034	0.02	0.001

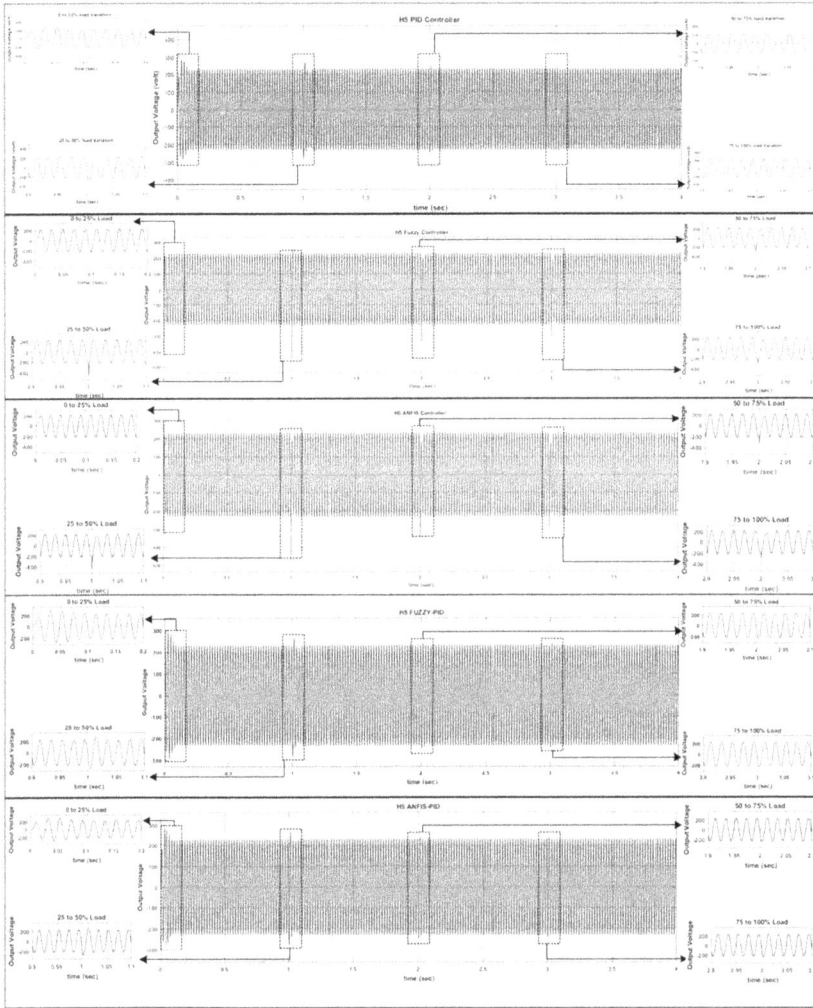

FIGURE 5.16 Load variation under operation for different control algorithms implemented in the H5 topology.

is increased; for the lower load, the settling time value is higher, and as higher load variation takes place, the settling time is just a few milliseconds.

In Figure 5.16, a complete analysis of the H5 topology is presented. The response to load variation with different control algorithms is described in the figure.

From Figure 5.16, it can be observed that the most variation is obtained for 25%–50% load variation in the case of fuzzy and ANFIS operation; for all the remaining, the variations are not too severe in amplitude and even the settling time is less. For 75%– 100% load variation, satisfactory results are obtained with very less variation instability and low settling time as well.

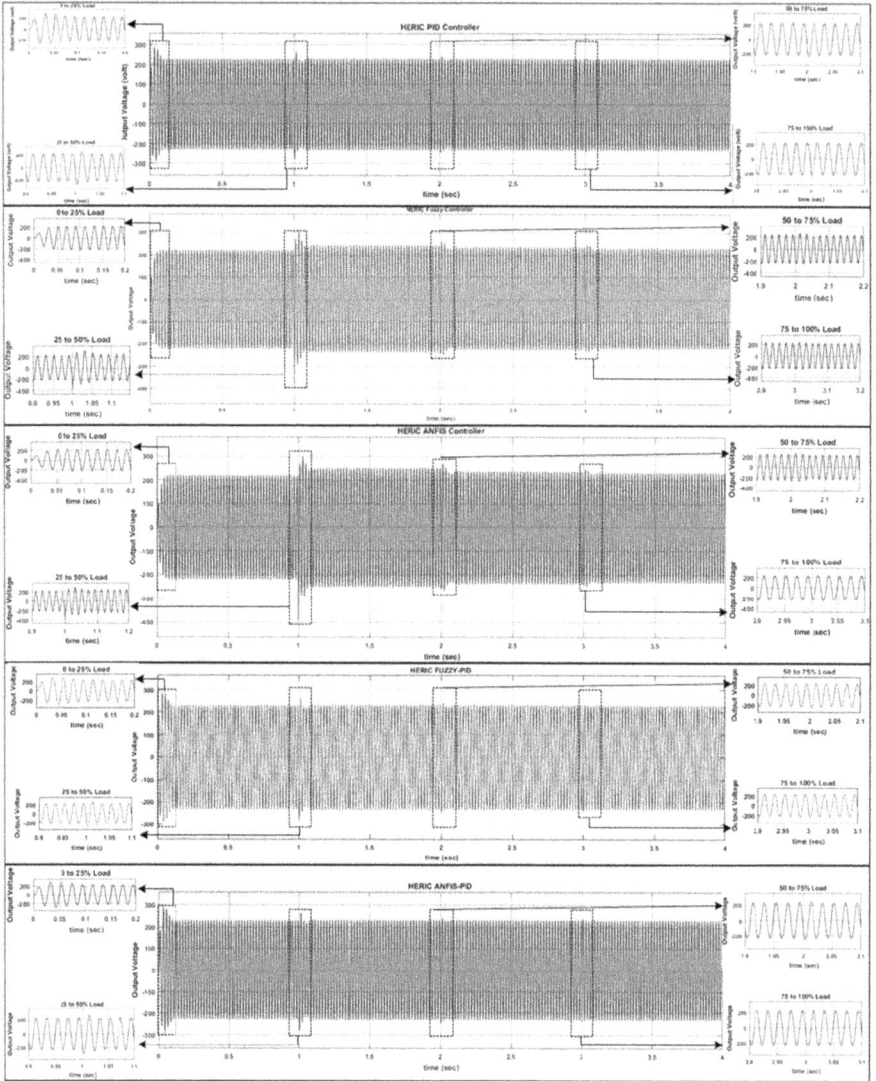

FIGURE 5.17 Load variation under operation for different control algorithms implemented in the HERIC topology.

In Figure 5.17, a complete analysis of HERIC topology is presented. The response to load variation with different control algorithms is described in the figure.

Like the HERIC, a larger variation is observed in magnitude for the 25%– 50% load variation in fuzzy and ANFIS control algorithms, as depicted in the figure above. The most satisfactory output is visible in the ANFIS-PID algorithm where the settling time is low, and the inverter tends to become stable faster.

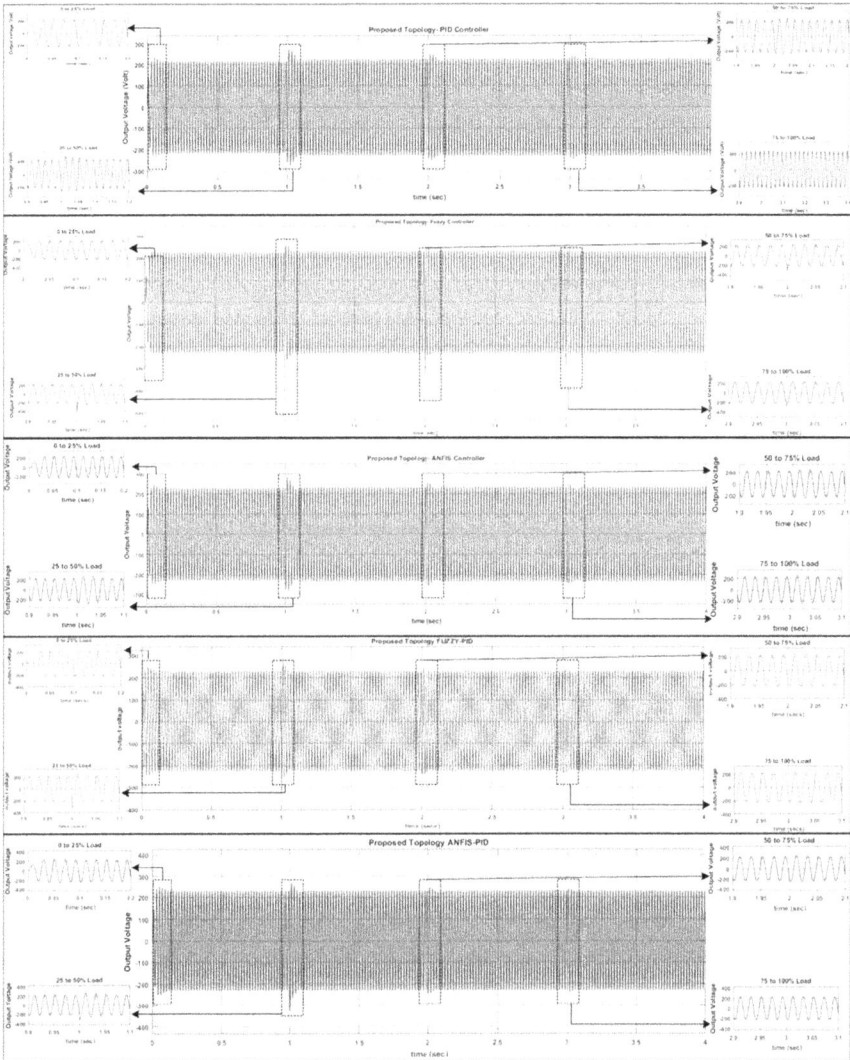

FIGURE 5.18 Load variation under operation for different control algorithms implemented in the proposed topology.

In Figure 5.18, a complete analysis of the proposed topology is presented. The response to load variation with different control algorithms is described in the figure.

In the above, the operation of the fuzzy algorithm has a variation of 25%–50% of load, whereas, for the remaining, the variation is significantly low. All the algorithms gave very satisfactory results when it comes to 75%– 100% load variation of the inverter.

The simulation in comparison with few previous works is presented in Table 5.6.

TABLE 5.6

Comparative Analysis of Different Parameters of Various Topologies with the Proposed Controller

	H5		HERIC		Proposed
Parameters	**Value [ref]**	**Simulated Result**	**Value [ref]**	**Simulated Result**	**Simulated Result**
THD	3.29% [47]	2.96%	3.3% [48]	2.93%	1.33%
Settling time	-NA-	0.001 seconds	-NA-	0.001 seconds	0.001 seconds

5.4.6 TOTAL HARMONIC DISTORTION

For the harmonic analysis, fast Fourier transformation (FFT) analysis is implemented because of its computational efficiency. For calculating harmonic distortion, FFT analysis is done, and fundamental and other components can also be isolated by it. For the calculation of THD, a sum of the root-mean-square (RMS) value of voltage along with second to Nth harmonics is done and squared, which is divided by the fundamental component. THD is generally considered regarding percentage.

$$\text{THD\% of fundamental} = \frac{Q_{\text{rmsdistorted}}}{Q_{\text{fundamental}}} \times 100 \tag{5.7}$$

THD is a characteristic of the waveform. Low THD during high load has a much more severe impact on the system than the high THD will have during low load. As per the international standards, IEEE 516 [49] allows the range of voltage distortion in THD to be up to 5%. In Table 5.7, it can be observed that by FFT analysis, THD is calculated and the THD value for all the different topologies of the inverter under different control algorithms are well under the range specified. A comparison with some of the previous research works is presented in Table 5.6. THD for the proposed topology reduces up to 1.33% when compared with the existing topologies. Even by the implementation of control on existing topologies, a more satisfactory result is obtained in the majority of the cases.

5.4.7 STABILITY ANALYSIS OF THE INVERTER

Stability analysis is performed using a Bode plot of a closed-loop system. The magnitude and the phase angle of the system are calculated in the case of the Bode plot analysis of the system. In the Bode plot, a few of the characteristics of the system such as gain margin, phase margin, gain crossover frequency, and phase crossover frequency is determined and is further taken into consideration while analyzing system stability.

The MATLAB linearization tool was implemented in the research for studying the different topologies' stability analysis when they are operated on dissimilar control algorithms. In Table 5.8, it is presented that for the H5 topology of the transformerless inverter, when operating with various control algorithms, the system remains stable as, by studying the Bode plot, it was found that the gain margin and phase

TABLE 5.7

THD and FFT Analysis of Different Topologies Using Various Algorithms

	TOPOLOGY		
Control	H5	HERIC	Proposed

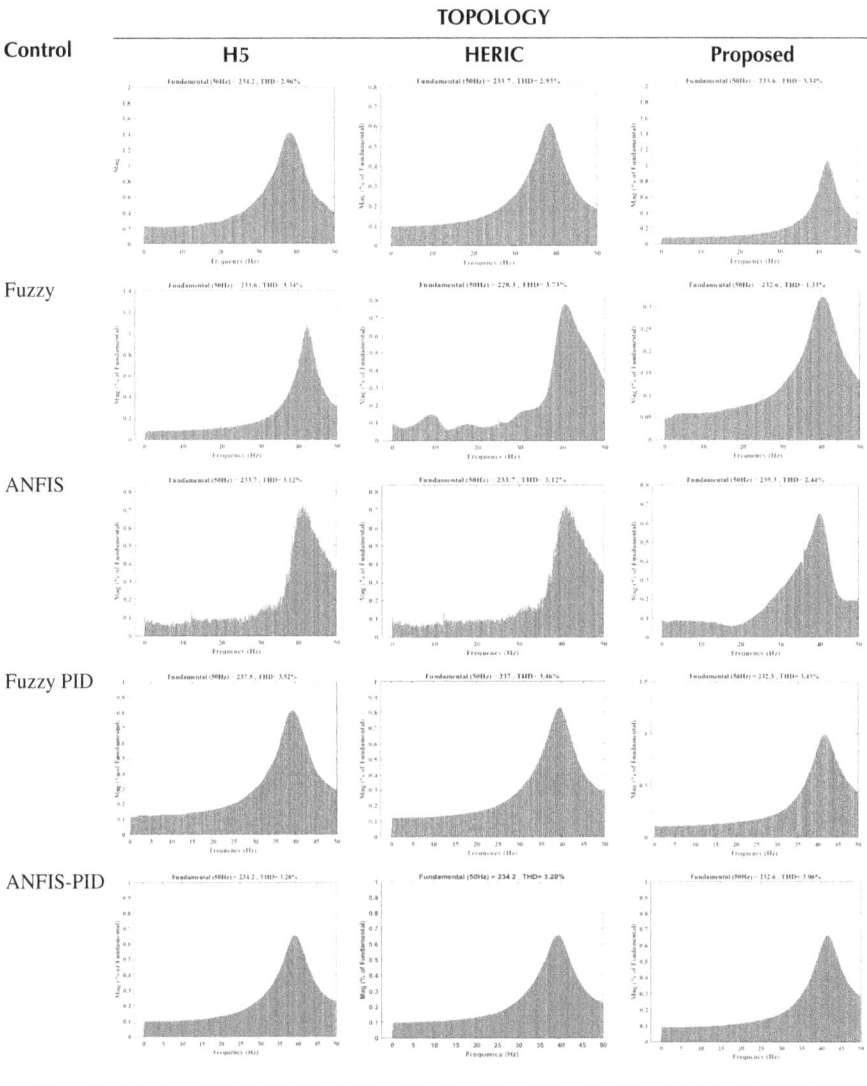

margin of all the topologies are in the same half [50]. The gain crossover frequency and phase crossover frequency are found using the plot illustrated in Figure 5.19.

In Table 5.9, it is presented that for the HERIC topology of the transformerless inverter when operating using various control algorithms, the system remains stable as by studying the Bode plot, it was found that the gain margin and phase margin of all the topologies are in the same half. The gain crossover frequency and phase crossover frequency are found using the plot illustrated in Figure 5.20.

TABLE 5.8

Bode Plot Characteristics of H5 Topology with Different Control Algorithms

	H5 Topology				
	PID	**Fuzzy**	**ANFIS**	**Fuzzy PID**	**ANFIS-PID**
Gain margin (G_m)	0.6003	1.4167	1.5	2.4342	0.0007012
Phase margin (P_m)	60.8492	46.2738	46.5	24.9995	90.0016
Gain crossover frequency (W_{gm})	0.001	0.001	0.001	0.001	1.7286
Phase crossover frequency (W_{pm})	0.0328	0.00714	0.007	0.0892	7.23e-11

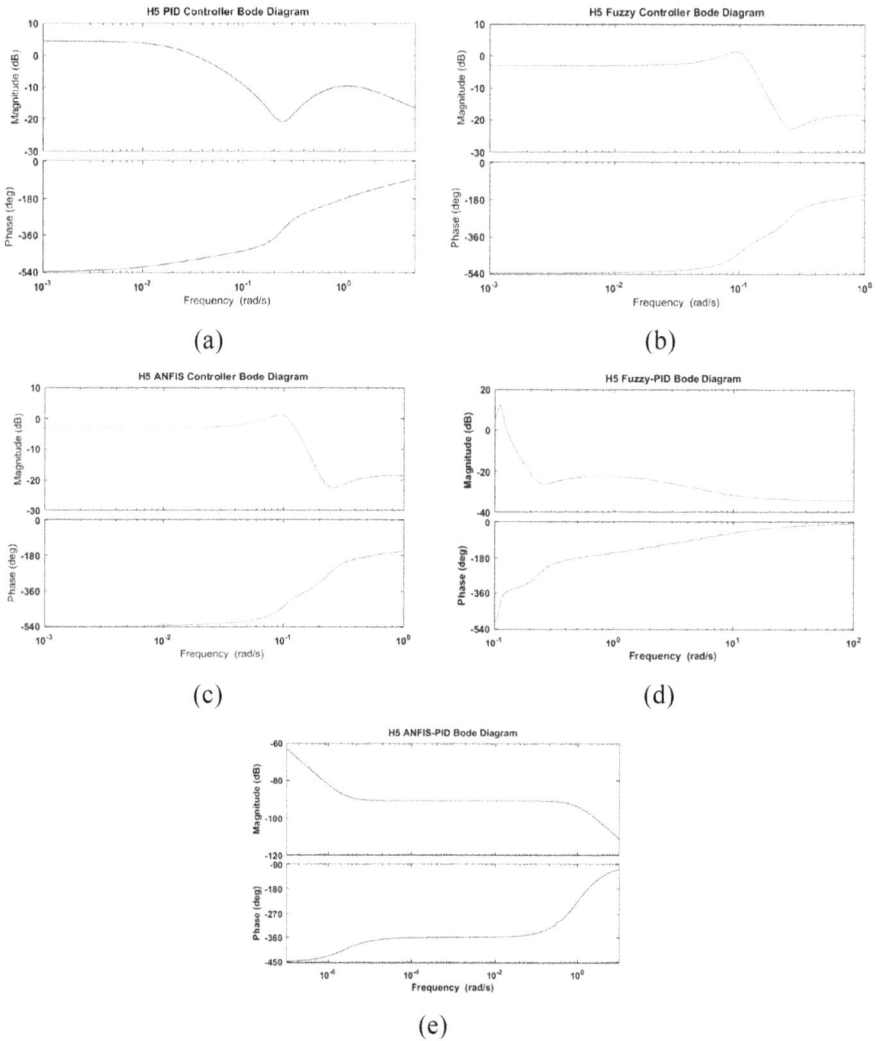

FIGURE 5.19   H5 topologies' Bode plot on different control algorithms.

TABLE 5.9
Bode Plot Characteristics of HERIC Topology with Different Control Algorithms

	HERIC Topology				
	PID	**Fuzzy**	**ANFIS**	**Fuzzy PID**	**ANFIS-PID**
Gain margin (G_m)	0.6265	1.0755e+05	1.6242e+07	1.1198e+05	7.2476e+04
Phase margin (P_m)	150.29	90.0011	90	90.0005	90.0008
Gain crossover frequency (W_{gm})	0	1.7331	1.7332	1.3	1
Phase crossover frequency (W_{pm})	0.1599	1.0528e-10	6.9731e-13	2.2393e-11	3.4606e-11

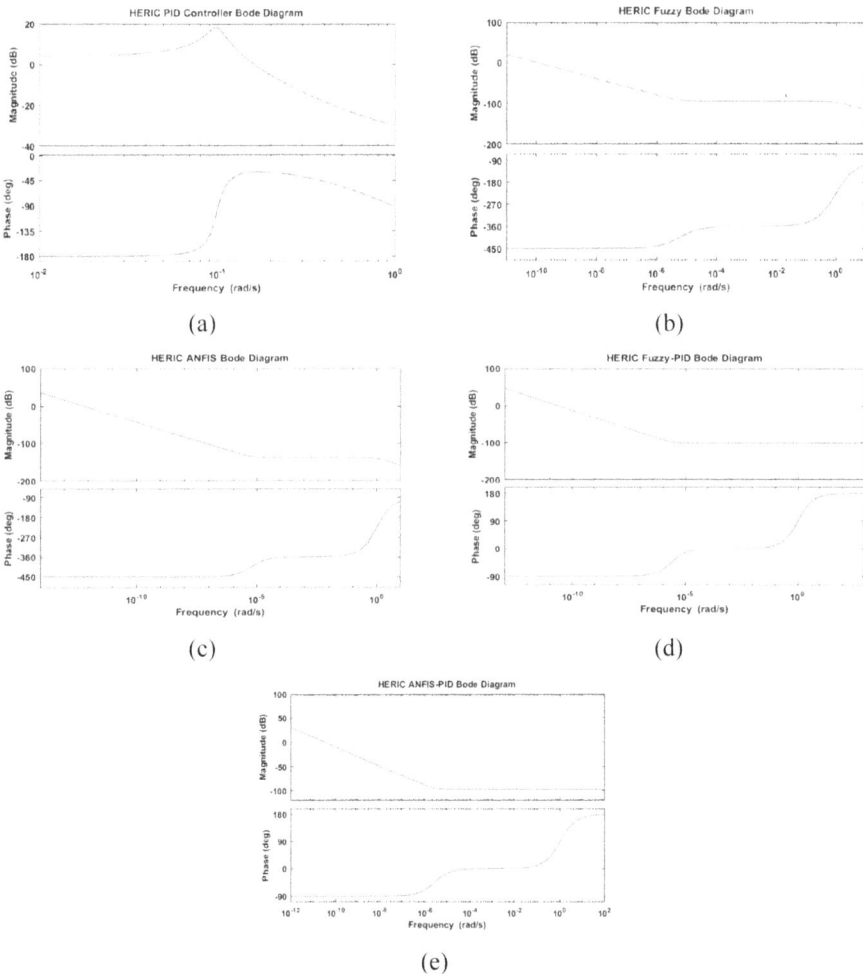

(a)

(b)

(c)

(d)

(e)

FIGURE 5.20 HERIC topologies' Bode plot on different control algorithms.

TABLE 5.10

Bode Plot Characteristics of the Proposed Topology with Different Control Algorithms

	Proposed Topology				
	PID	**Fuzzy**	**ANFIS**	**Fuzzy PID**	**ANFIS-PID**
Gain margin (G_m)	4.8366e+06	228.1469	3.5096e+04	1.54e+05	251.451
Phase margin (P_m)	168.52	136.2342	96.8572	121.53	91.169
Gain crossover frequency (W_{gm})	1.27e+03	3.966	3.9661	3.96	3.9668
Phase crossover frequency (W_{pm})	0.1	0.0033	3.015e-05	5.9e-06	0.0042

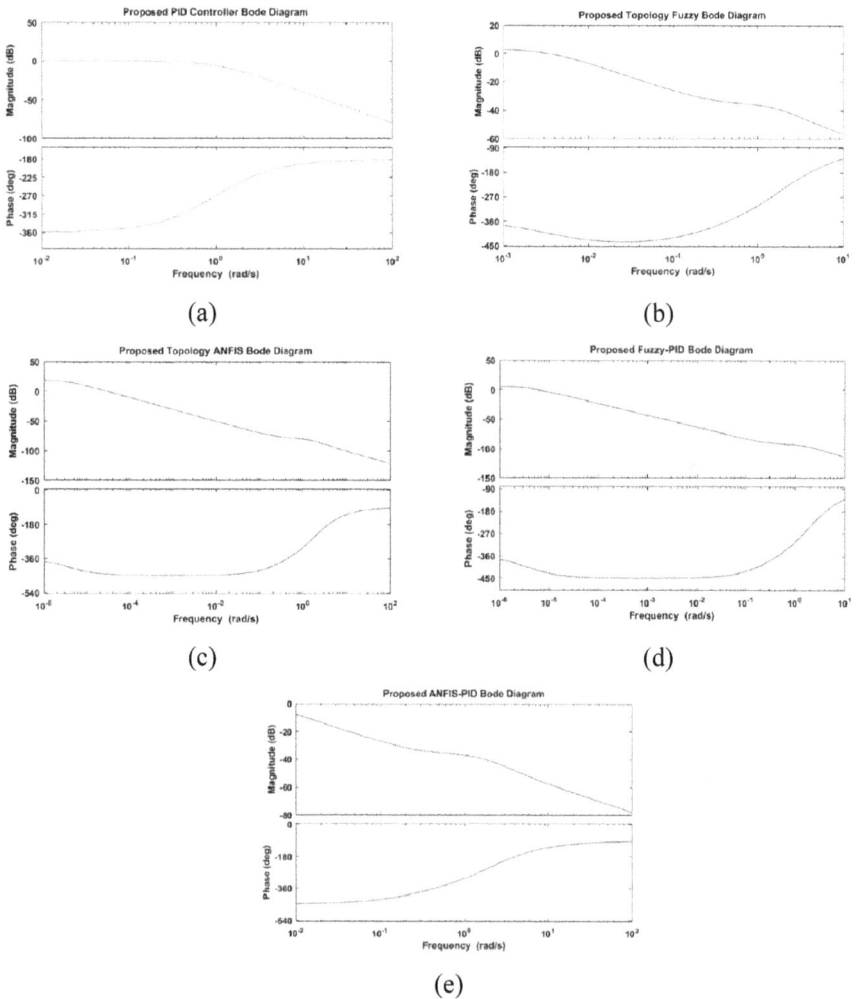

(a)

(b)

(c)

(d)

(e)

FIGURE 5.21 Proposed topologies' Bode plot on different control algorithms.

In Table 5.10, it is presented that, for the proposed topology, when operating with various control algorithms, the system remains stable as by studying the Bode plot, it was found that the gain margin and phase margin of all the topologies are in the same half. The gain crossover frequency and phase crossover frequency are found in the plot illustrated in Figure 5.21.

5.5 SUMMARY

The work presented in this chapter is about the control strategies of inverters with a novel control strategy and with a new topology. The H5 and HERIC topologies of inverters are described, and a novel inverter topology has been developed. The popular switching strategies are summarized with an emphasis on carrier-based PWM. The topic of voltage balancing without any auxiliary power circuit is explained, as well as its control algorithm for all the topologies was implemented. Furthermore, the dynamic model of the H5, HERIC, and proposed topologies were developed. Both the control part and load balancing have been realized in MATLAB-Simulink. The control strategies have been designed for their subsequent implementation into the real-time control system. Both the balancing algorithm and the performed analysis on all the topologies using various control techniques like PID, FUZZY, FUZZY PID, ANFIS, and ANFIS-PID have been verified using simulation results that have also been presented. The performance parameters like stability analysis, settling time, noise, etc. are evaluated and compared with the existing work. It is found that the leakage current for H5 and HERIC topology with the new control strategy is reduced to 24 and 53.2 mA, respectively, as compared to previous work reported. The reduction in THD is also reduced to 1.3% from the previously reported 3.29%. It is evident from the results that the proposed work has improved the performance parameters.

5.6 QUESTIONS

1. What is the standalone operation of inverters?
2. How is it different from grid-connected inverters?
3. What are the design criteria for standalone inverters?
4. What are power quality factors?
5. How is stability analysis performed in standalone inverter control?

REFERENCES

[1] H. Xiao, S. Xie, Y. Chen, and R. Huang, "An optimized transformerless photovoltaic grid-connected inverter," *IEEE Trans. Ind. Electron.*, vol. 58, no. 5, pp. 1887–1895, 2011, doi: 10.1109/TIE.2010.2054056.
[2] I. Patrao, E. Figueres, F. González-Espín, and G. Garcerá, "Transformerless topologies for grid-connected single-phase photovoltaic inverters," *Renew. Sustain. Energy Rev.*, vol. 15, no. 7, pp. 3423–3431, Sep. 2011, doi: 10.1016/j.rser.2011.03.034.
[3] E. Gubía, P. Sanchis, A. Ursúa, J. López, and L. Marroyo, "Ground currents in single-phase transformerless photovoltaic systems," *Prog. Photovoltaics Res. Appl.*, vol. 15, no. 7, pp. 629–650, Nov. 2007, doi: 10.1002/pip.761.

[4] J. R. Rodríguez, E. L. Moreno-Goytia, V. V. Rebollar, L. E. Ugalde, and N. M. Salgado-Herrera, "Step-up transformerless seven-level DC-AC hybrid topology for interconnection of renewable-based dc sources to microgrids," *Electr. Power Components Syst.*, vol. 42, no. 16, pp. 1792–1801, 2014, doi: 10.1080/15325008.2014.949910.

[5] T. K. S. Freddy, N. A. Rahim, W. Hew, and H. S. Che, "Comparison and analysis of single-phase transformerless grid-connected PV inverters," *IEEE Trans. Power Electron.*, vol. 29, no. 10, pp. 5358–5369, Oct. 2014, doi: 10.1109/TPEL.2013.2294953.

[6] M. Azri, N. A. Rahim, and M. F. M. Elias, "Transformerless DC/AC converter for grid-connected PV power generation system," *Arab. J. Sci. Eng.*, vol. 39, no. 11, pp. 7945–7956, 2014, doi: 10.1007/s13369-014-1308-z.

[7] J. M. Sosa, G. Escobar, M. A. Juarez, and A. A. Valdez, "H5-HERIC based transformerless multilevel inverter for single-phase grid connected PV systems," in *IECON 2015*, 2015, pp. 1026–1031.

[8] S. Hu, C. Li, and W. Li, "Enhanced HERIC based transformerless inverter with hybrid clamping cell for leakage current elimination," in *Energy Conversion Congress and Exposition (ECCE)*, 2015, pp. 5337–5341.

[9] M. Islam, S. Mekhilef, and M. Hasan, "Single phase transformerless inverter topologies for grid-tied photovoltaic system: a review," *Renew. Sustain. Energy Rev.*, vol. 45, pp. 69–86, May 2015, doi: 10.1016/j.rser.2015.01.009.

[10] X. Fu and S. Li, "Control of single-phase grid-connected converters with LCL filters using recurrent neural network and conventional control methods," *IEEE Trans. Power Electron.*, vol. 31, no. 7, pp. 5354–5364, 2016, doi: 10.1109/TPEL.2015.2490200.

[11] A. Datta, G. Bhattacharya, D. Mukherjee, and H. Saha, "Modelling and simulation-based performance study of a transformerless single-stage grid-connected photovoltaic system in Indian ambient conditions," *Int. J. Ambient Energy*, vol. 37, no. 2, pp. 172–183, 2016, doi: 10.1080/01430750.2014.915885.

[12] P. K. Pranav and P. N. Korde, "Double tuned resonant filter used transformer less grid connected photovoltaic system," *2016 in International Conference on Energy Efficient Technologies for Sustainability ICEETS*, IEEE, pp. 533–537, 2016, doi: 10.1109/ICEETS.2016.7583811.

[13] X. Fu and S. Li, "A novel neural network vector control for single-phase grid-connected converters with L, LC and LCL filters," *Energies*, vol. 9, no. 5, 2016, doi: 10.3390/en9050328.

[14] H. Li, Y. Zeng, and T. Q. Zheng, "A novel H5-D topology for transformerless photovoltaic grid-connected inverter application," in *2016 IEEE 8th International Power Electronics and Motion Control Conference (IPEMC-ECCE)*, Asia, IEEE, pp.731–735.

[15] S. A. Khan, Y. Guo, and J. Zhu, "A high efficiency transformerless PV grid-connected inverter with leakage current suppression," in *9th International Conference on Electrical and Computer Engineering, (ICECE)*, IEEE, 2016, pp. 190–193.

[16] P. Somani, "Design of HERIC configuration based grid connected single phase transformer less ;photovoltaic inverter," in *International Conference on Electrical, Electronics, and Optimization Techniques (ICEEOT)*, IEEE, 2016, pp. 892–896.

[17] M. Sun, M. Zhao, K. Qu, F. Li, L. Mao, and M. Feng, "Novel single-phase transformerless inverter based on the freewheeling loop separation for PV systems," in *2016 IEEE 8th International Power Electronics and Motion Control Conference (IPEMC-ECCE)*, Asia, IEEE, pp. 1527–1531.

[18] L. Xiao, Hua F.; Zhang, "A zero-voltage-transition HERIC-type transformerless photovoltaic grid-connected inverter," *IEEE Trans. Ind. Electron.*, vol. 64, no. 2, pp. 1222–1232, 2017.

[19] Z. Ahmad and S. N. Singh, "Comparative analysis of single phase transformerless inverter topologies for grid connected PV system," *Sol. Energy*, vol. 149, pp. 245–271, 2017, doi: 10.1016/j.solener.2017.03.080.

[20] A. I. Maswood, "A PWM voltage source inverter with pi controller for instantaneous motor current control," in *International Conference on Power Electronics and Drive Systems*, IEEE, 1995, no. 95, pp. 834–837.

[21] J. Selvaraj, N. A. Rahim, and C. Krismadinata, "Digital PI current control for grid connected PV inverter," in *3rd IEEE Conference on Industrial Electronics and Applications (ICIEA)*, pp. 742–746, 2008, doi: 10.1109/ICIEA.2008.4582614.

[22] S. M. Ayob, Z. Salam, N. A. Azli, and M. E. Elbuluk, "Control of a single phase inverter using fuzzy logic," in *Conference Record of the IEEE Industry Applications Society Annual Meeting Electr. Eng.*, IEEE, 2009, pp. (1–6).

[23] E. Villanueva, P. Correa, J. Rodríguez, and M. Pacas, "Control of a single-phase cascaded H-bridge multilevel inverter for grid-connected photovoltaic systems," *IEEE Trans. Ind. Electron.*, vol. 56, no. 11, pp. 4399–4406, 2009.

[24] G. Shen, X. Zhu, and D. Xu, "A new feedback method for PR current control of LCL filter based grid connected inverter," *IEEE Trans. Ind. Electron.*, vol. 57, no. 6, pp. 2033–2041, 2010.

[25] A. El Khateb, N. A. Rahim, and J. Selvaraj, "Optimized PID controller for both single phase inverter and MPPT SEPIC DC/DC converter of PV module," in *2011 IEEE International Electric Machines and Drives Conference IEMDC 2011*, pp. 1036–1041, 2011, doi: 10.1109/IEMDC.2011.5994743.

[26] T. Kerekes, R. Teodorescu, P. Rodríguez, G. Vázquez, and E. Aldabas, "A new high-efficiency single-phase transformerless PV inverter topology," *IEEE Trans. Ind. Electron.*, vol. 58, no. 1, pp. 184–191, 2011.

[27] G. Buticchi, D. Barater, E. Lorenzani, and G. Franceschini, "Digital control of actual grid-connected converters for ground leakage current reduction in PV transformerless systems," *IEEE Trans. Ind. Informatics*, vol. 8, no. 3, pp. 563–572, 2012, doi: 10.1109/TII.2012.2192284.

[28] Y. Yang, F. Blaabjerg, and H. Wang, "Low-voltage ride-through of single-phase transformerless photovoltaic inverters," *IEEE Trans. Ind. Appl.*, vol. 50, no. 3, pp. 1942–1952, 2014, doi: 10.1109/TIA.2013.2282966.

[29] L. Hassaine, E. Olias, J. Quintero, and V. Salas, "Overview of power inverter topologies and control structures for grid connected photovoltaic systems," *Renew. Sustain. Energy Rev.*, vol. 30, pp. 796–807, 2014, doi: 10.1016/j.rser.2013.11.005.

[30] R. J. Wai and Y. K. Liu, "Design of fuzzy neural network control for single-stage boost inverter," in *Proceeding-International Conference on Machine Learning and Cybernetics (ICMLC)*, vol. 2, no. 12, pp. 560–565, 2015, doi: 10.1109/ICMLC.2015.7340615.

[31] I. Sefa, N. Altin, S. Ozdemir, and O. Kaplan, "Fuzzy PI controlled inverter for grid interactive renewable energy systems," *IET Renew. Power Gener.*, vol. 9, pp. 729–738, 2015, doi: 10.1049/iet-rpg.2014.0404.

[32] R. K. Ahuja, T. Maity, and S. Kakkar, "Control of active and reactive power of grid connected inverter using adaptive network based fuzzy inference system (ANFIS)," in *2016 IEEE 7th Power India International Conference (PIICON)*, 2016, pp. 0–4.

[33] M. Rajeev and V. Agarwal, "Closed loop control of novel transformer-less inverter topology for single phase grid connected photovoltaic system," *(PVSC)*, IEEE, 2016, pp. 1–5.

[34] T. K. Mohan and S. F. Mohammed, "A neuro-fuzzy controller for multilevel renewable energy system," *Int. Conf. Electr. Electron. Optim. Tech. ICEEOT 2016*, pp. 4120–4123, 2016, doi: 10.1109/ICEEOT.2016.7755491.

[35] S. Chakraborty, M. M. Hasan, and M. Abdur Razzak, "Transformer-less single-phase grid-tie photovoltaic inverter topologies for residential application with various filter circuits," *Renew. Sustain. Energy Rev.*, vol. 72, no. October, pp. 1152–1166, May 2017, doi: 10.1016/j.rser.2016.10.032.

[36] M. Islam and S. Mekhilef, "Efficient transformerless MOSFET inverter for a grid-tied photovoltaic system," *IEEE Trans. Power Electron.*, vol. 31, no. 9, pp. 6305–6316, Sep. 2016, doi: 10.1109/TPEL.2015.2501022.

[37] J. Fei and Y. Zhu, "Adaptive fuzzy sliding control of single-phase PV grid-connected inverter," *2017 IEEE Int. Conf. Mechatronics Autom.*, pp. 1233–1238, 2017, doi: 10.1109/ICMA.2017.8015993.

[38] Q. Trinh, F. H. Choo, and P. Wang, "Control strategy to eliminate impact of voltage measurement errors on grid current performance of three-phase grid connected inverters," *IEEE Trans. Ind. Electron.*, vol. 64, no. 9, pp. 7508–7519, 2017.

[39] H. Anantwar, B. R. Lakshmikantha, and S. Sundar, "Fuzzy self tuning PI controller based inverter control for voltage regulation in off-grid hybrid power system," *Energy Procedia*, vol. 117, pp. 409–416, 2017, doi: 10.1016/j.egypro.2017.05.160.

[40] M. A. Khan, A. Haque, V. S. B. Kurukuru, H. Wang, and F. Blaabjerg, "Stand-alone operation of distributed generation systems with improved harmonic elimination scheme," *IEEE J. Emerg. Sel. Top. Power Electron.*, vol. 9, no. 6, pp. 6924–6934, Dec. 2021, doi: 10.1109/JESTPE.2021.3084737.

[41] M. A. Khan, A. Haque, and V. S. B. Kurukuru, "Performance assessment of stand-alone transformerless inverters," *Int. Trans. Electr. Energy Syst.*, vol. 30, no. 1, pp. 1–20, Jan. 2020, doi: 10.1002/2050–7038.12156.

[42] M. A. Khan, A. Haque, and V. S. B. Kurukuru, "Intelligent control of a novel transformerless inverter topology for photovoltaic applications," *Electr. Eng.*, vol. 102, no. 2, pp. 627–641, Jun. 2020, doi: 10.1007/s00202-019-00899-2.

[43] Goverment of India, "National Electrical Code," 2011.

[44] M. A. Khan, A. Haque, V. S. B. Kurukuru, and M. Saad, "Islanding detection techniques for grid-connected photovoltaic systems-a review," *Renew. Sustain. Energy Rev.*, vol. 154, p. 111854, Feb. 2022, doi: 10.1016/j.rser.2021.111854.

[45] M. A. Khan and A. Sangwongwanich, "Control strategy for grid-connected solar inverter for IEC standards," in *Reliability of Power Electronics Converters for Solar Photovoltaic Applications*, Institution of Engineering and Technology, 2021, pp. 141–188.

[46] H. Markiewicz and A. Klajn, "Power quality application guide-voltage disturbances standard EN 50160," *Copp. Dev. Assoc.*, vol. 5.4.2, pp. 4–11, 2004.

[47] H. Cao, "A novel inverter topology for single-phase," *Act. Passiv. Electron. Components*, vol. 2016, 2016.

[48] S. A. A. Zaid and A. M. Kassem, "Review, analysis and improving the utilization factor of a PV-grid connected system via HERIC transformerless approach," *Renew. Sustain. Energy Rev.*, vol. 73, no. September 2015, pp. 1061–1069, Jun. 2017, doi: 10.1016/j.rser.2017.02.025.

[49] "IEEE guide for maintenance methods on energized power lines," *IEEE Std 516-2003 (Revision IEEE Std 516–1995)*, 2003.

[50] I. J. Nagrath and M. Gopal, *Control System Engineering*, Fifth. New Delhi: New Age International Publication, 2017.

6 Grid Connected Operation of PV Inverter

6.1 INTRODUCTION

Photovoltaic (PV) inverters are one of the key components in a PV power plant. A PV inverter is responsible for transferring PV energy to the grid in a highly efficient and robust way, grid synchronization, anti-islanding, voltage, and current protections to protect PV installation, maximum power extraction from PV modules and etc. [1,2]. Among the various types of PV system configuration concepts, the multi-string inverter is relatively a new configuration that contains more than one DC–DC converter for PV strings eliminating the negative effects of partial shading, and one inverter to transfer PV energy to the grid which makes it a hybrid solution for intermediate power level between residential and utility-scale PV plants by containing advantageous parts of string inverter and central inverter [2].

By utilizing new-generation power semiconductors, inverter efficiencies have reached very high efficiencies up to 98.2% for ABB TRIO-27.6-TL transformerless multistring inverter with new-generation IGBT power semiconductors [3]. In addition to existing power semiconductor technology, SiC-based power MOSFETs have been commercialized in recent years. Lower ton and toff time of SiC MOSFETs lead lower switching loss so that higher efficiency up to 99.5% [4] and reduction of heat-sink volumes, also enables utilization higher switching frequency so that reduction passive component size. Based on these features, SiC power MOSFETs can dominate the power semiconductor market with the price reduction in next years.

6.2 OVERVIEW OF PV SYSTEMS

Photovoltaic system consists of some various components such as PV modules, junction box, solar inverters, mounting, cabling, and in some installations battery system, battery charger, solar tracker, maximum power point tracker (MPPT), transformer station and so on [5]. PV modules are used in order to transform solar power to electricity, and the serial connections of modules constitute a PV string. Most of modules consist of generally 60–72 cells and generate a DC voltage of 30–40 V with a power range of 160–300 W. With the serial connection of PV modules, PV string generate a DC voltage between 400 and 950 V in most applications. A number of PV strings are paralleled depending to the inverter power rating. Those PV string cables goes to a combining box named as junction box which includes connection cables of PV strings, fuses for each string, DC disconnector, surge arrestor for lightning protection. PV inverter is responsible for transferring the combined power of PV strings into grid in an efficient, robust way [6]. There are some regulations indicated in international standards for PV inverters which will be explained in this chapter.

DOI: 10.1201/9781003257189-6

Also, in some installations especially where solar power station is not connected to grid infrastructure, battery system and battery chargers are used in order to maintain providing electricity to loads. In addition, those power stations with battery system can be used to support active power to grid in cloudy weather and at the night hour when the interconnected system needs active power [7].

Photovoltaic energy has a wide range of applications starting from a few hundreds of watts for a small system to more than a gigawatt for large utility-scale PV plants. Because of this wide power range, different PV system configurations is used for different power levels of PV installations. Each PV configuration concept has its own advantageous and disadvantageous features which will be discussed in this chapter.

6.2.1 Photovoltaic Module Characteristics

Photovoltaic panel characteristics change with conditions such as solar irradiance and ambient temperature. An electrical circuit of a multi-string PV system based on a static model is given in Figure 6.1. Multi-string PV system is composed of individual PV modules so that it can be observed as the electrical circuit of a PV module.

As can be seen in Figure 6.1 PV module is represented as a photo-current source where R_s and R_{sh} are equivalent series and shunt resistance of PV string. Also, cabling R_{cab} and L_{cab} serial resistance and inductance values are added to the static model. Although the static model can lead to sufficient resemblance with real cases for steady state DC loading applications, for transient stages and pulsed current loading applications an alternative model named in the literature as a dynamic model which is shown in Figure 6.2 is used [8]. This dynamic model gives many accurate

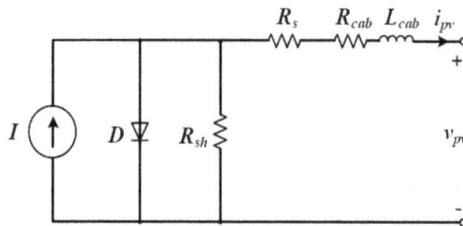

FIGURE 6.1   Equivalent circuit of multi-string PV system for a static model.

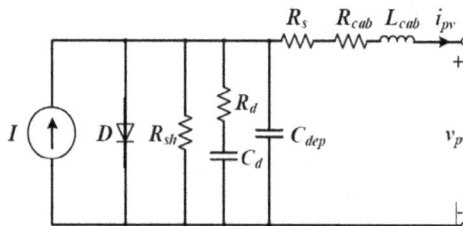

FIGURE 6.2   Equivalent circuit of multi-string PV system for the dynamic model.

results compared to static models therefore for simulations of the MPPT converter dynamic model should be used.

As stated, earlier PV module output power is significantly dependent on ambient temperature and solar irradiance. As it can be foreseen, increasing solar irradiance results in increase of output power and ambient temperature increase effects output power negatively. Furthermore, as the PV panel output current increases, PV panel output voltage decreases non-linearly. Also, with the increasing solar irradiance for curves from brown to red, open circuit panel voltage and short circuit panel current hence maximum output power increases. Because of the non-linear change of PV panel output current and voltage, change of PV panel output power curve is expected to be a hill form with change of loading conditions [9–11].

Moreover, panel output power reaches a maximum power point which occurs at a fixed loading condition and this loading point changes with solar irradiance and ambient temperature. Maximum power point tracking duty which can be done by a separate DC–DC converter or by inverter itself, aims to extract the maximum power from PV string, by changing the loading conditions. For maximum power point tracking, different algorithms exist such as perturb-observe, incremental conductance methods and some hybrid forms [12,13]. Due to the fact that a separately designed MPPT converter is used in this research work, maximum power point tracking subject will not be detailed.

6.2.2 Photovoltaic System Requirements

In order to establish a PV system in an efficient and robust way and to keep power quality of transferred energy at desired standards, there are some expected requirements for PV systems. To be able to get maximum use of installed PV strings, maximum power point tracking systems are used. According to the mission profile and rated power of the PV system, various PV configuration types can be used as well as inverter types and MPPT control algorithms. Apart from that, there are some regulations for the grid side requirements such as total harmonic distortion value of a lower level of 5% [14].

For high power rating PV systems, such as larger than hundreds kW, it is expected from PV system to support grid by providing or consuming reactive power and other necessary precautions for stabilizing grid voltage. In addition to these general requirements, there are some more requirements changing by countries. Since the PV technology is more expensive compared to other energy sources such as wind energy there is a strong urge to increase the efficiency of PV system. In order to increase the efficiency, transformerless PV inverters are allowed and gain popularity in the European market [15]. On the other hand, absence of galvanic isolation can cause a leakage current which can result in safety issues, so that reduction of leakage current is required in most cases [16,17].

Lastly reliability of PV system is also important for energy production to be not interrupted, since a possible failure can cause repairing costs as well as money loss due to idle system. Temperature intolerance is one of the most crucial factors that affects reliability of the system, due to the fact that exposure to sun and small inverter housings. High temperature results in degradation of some components of inverter. In order for prevention of failure due to temperature, inverter cooling mechanism must be well-designed and components with high temperature grade must be selected.

6.2.3 PHOTOVOLTAIC SYSTEM CONFIGURATION TYPES

Installed power of PV system installations changes from a couple of hundred Ws to MWs. Because of this large range of variety, for different power levels, different system structures are used. These system structures can be composed under four main configuration types [18]. As it is shown in Figure 6.3 for small system use, power level up to 300 W which is the average power of a PV panel, module inverter is used. For residential use, power level between 1 kW and 10 kW string inverter, for commercial and residential uses, power between 10 kW and 30 kW multi-string inverter, for commercial-utility scale uses, power level larger than 30 kW central inverter is used. These power levels which define the system configuration types are vague numbers and these configuration types can be used for different power levels, by considering advantages and disadvantages of each configuration type and system needs, in terms of efficiency, price, robustness, and partial shading conditions.

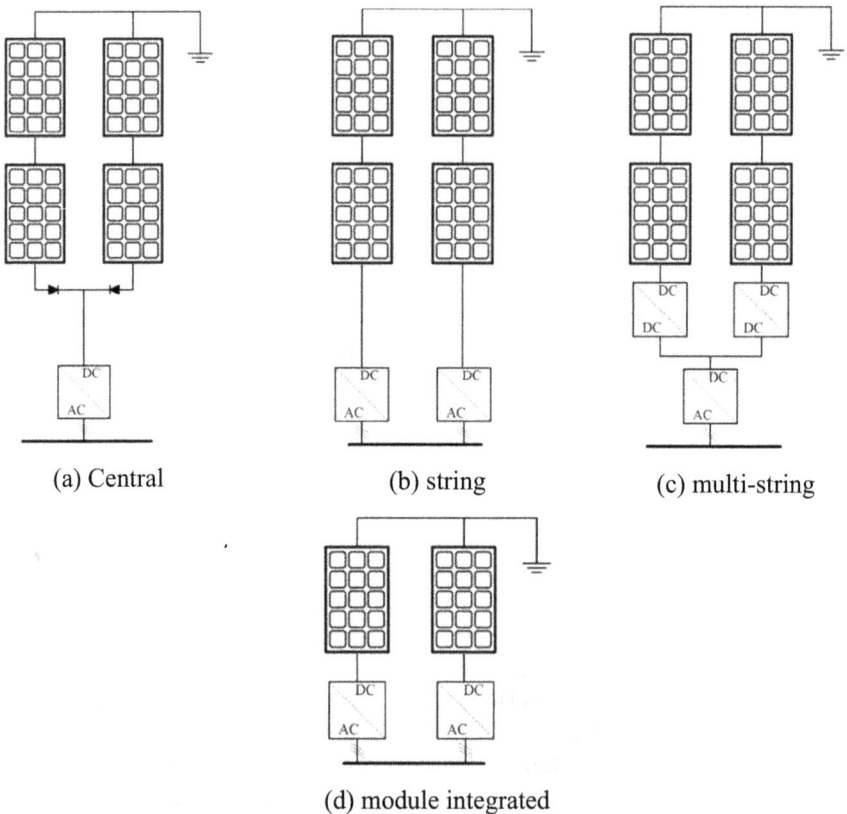

(a) Central   (b) string   (c) multi-string

(d) module integrated

FIGURE 6.3 Different PV system configuration types based on the power demanded. (a) Central inverter, (b) String inverter, (c) multi-string inverter, (d) module integrated inverter.

Module inverters are generally used for small PV installations. Since for each PV module a separate inverter is used in this configuration type, maximum power point tracking for this type of configuration is optimized and in the case of a malfunction, missing power is minimum during the maintenance and repair time. The disadvantages of this configuration type are mainly price and efficiency. Price/Watt for this type of inverter is higher compared to other types of inverters and efficiency values are lower compared to higher power rating inverters. Because of these listed disadvantages, those type of inverters are generally used in small systems. String inverters are used as connected to a single PV string, so that each string maximum power point is tracked individually and this results in better MPPT. String inverters can be connected to single or three phase grids. For the case of PV string voltage is not enough, a DC–DC converter is used to increase the DC link voltage. This DC–DC converter could be a boost type or a HF-link transformer-based converter. If the PV string voltage is enough and there is no need for galvanic isolation, PV inverter can be single stage, by taking necessary precautions for common mode currents. Also, there is no need for string diodes, so that string diode losses can be eliminated. By taking into consideration these conditions, string inverters are advantageous in many ways especially in partial shading conditions. Multi-string inverter is a hybrid solution which combines the advantages of central inverters, in terms of low cost and string inverter, in terms of individual MPP tracking for each string. For each PV string, there is a separate DC–DC converter in order to increase the voltage level and fulfill MPPT duty. Outputs of DC–DC converters are paralleled and transferred to grid by a single inverter. Also like in the string inverter case, a transformerless type can be adapted. Central inverters are generally used in large scale utility systems by combining PV strings through string diodes. Central inverters are single stage inverters without a DC–DC converter. Since all of the PV strings are paralleled, strings do not operate in the exact maximum power point. Also due to string diodes and very long cabling, additional losses occur. Because of the absence of DC–DC converter, PV modules must be connected such that, output voltage of PV string reaches enough voltage for DC link mostly between 550 and 850 V. Another disadvantage of central inverter is that a very large power source is dependent on a single inverter, and in the case of a malfunction, money loss is huge during the repair time. Despite these disadvantages, central inverters have a very high efficiency value up to 99% and very low Price/Watt.

6.2.4   SiC Based MOSFETs

The occurrence of wide bandgap switching devices, namely as SiC and GaN base material MOSFETs, enabled higher efficiency and higher power density in many power electronic devices. In the recent years, SiC technology in switching devices has been commercialized. SiC switching devices have lower turn-on and turn-off times compared to traditional Si-based switching elements. Also, on-state resistance for the fixed semiconductor area is relatively smaller in SiC devices. These conditions have resulted in better efficiency, increase of switching frequency, hence increase of power density in the area of power electronics application. Although switching frequency increase is not needed in all applications such as low voltage motor drives, since the

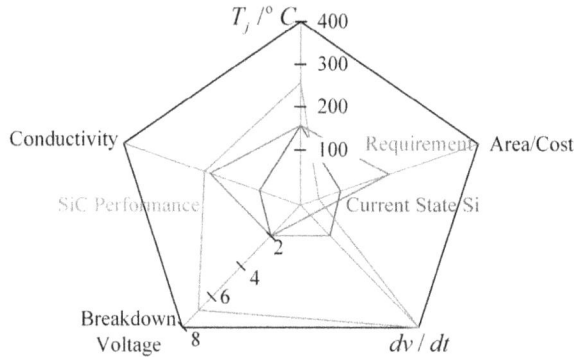

FIGURE 6.4 Radar chart comparison of Si and SiC-based switching devices in terms of $T_j/^\circ C$, *Area/cost*, *dv/dt*, breakdown voltage and conductivity [19].

motor inductance does enough filtering, higher efficiency for the same switching frequency is the advantage of SiC MOSFET use.

The radar chart comparison of traditional Si-based switching devices and newly developed SiC-based switching devices is given Figure 6.4, in terms of junction temperature $T_j/^\circ C$, Area/cost, *dv/dt*, breakdown voltage and conductivity.

As shown in Figure 6.4, SiC MOSFETs show better performance, in terms of every aspect, except the price which is expected to drop in the next years. Apart from the efficiency and power density due to switching frequency increase, junction temperature limitation of SiC MOSFETs is larger than 175°C, which is around 125°C for Si based semiconductors for the same voltage and current rating. This rise in the limitation of junction temperature, results in shrinking of heat-sink volume independent from efficiency value. Market share of wide bandgap devices of SiC- and GaN-based semiconductors is shown in Figure 6.5, by power rating and operation frequency.

As shown in Figure 6.5, use of SiC-based switching devices starts from 10 kHz operation frequency and power ratings up to 100 kVA, mainly in the applications of PV inverter, small pumps, adjustable speed drives, and automotive. In addition, with the newly developed commercial SiC devices, upper limit of power rating tends to increase.

6.2.5   VOLTAGE SOURCE THREE-PHASE INVERTER

Three-phase voltage source inverter consists of three half bridge switches and each half bridge switch is used to generate sinusoidal voltage waveform for each phase. Power bridge of three-phase voltage source inverter can be seen in Figure 6.6.

When each half bridge is switched according to the selected PWM method, unfiltered output of power bridge occurs as a pulse width modulated voltage waveform. In order to generate sinusoidal voltage waveform and to be able to transfer power to grid in a controlled way, a filter is needed. Single line diagram with an LCL filter is presented in Figure 6.7.

In order to transfer power to grid in a stable way, line impedance is needed since power transfer is done by changing phase and magnitude difference between inverter

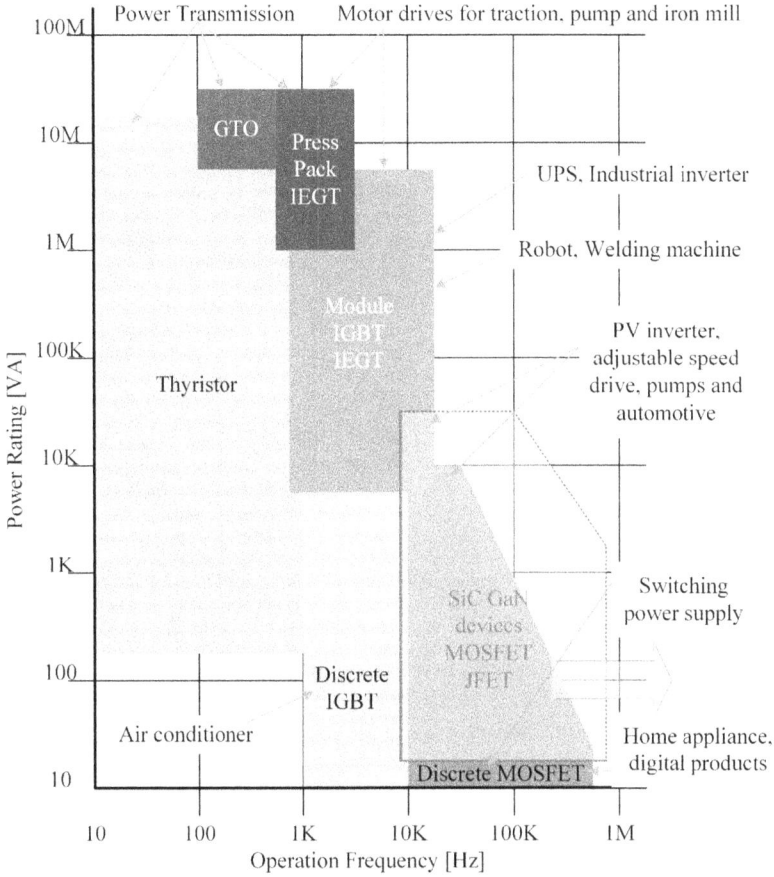

FIGURE 6.5 Market share of switching devices [20].

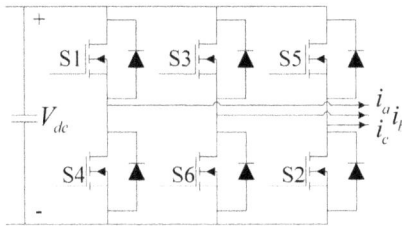

FIGURE 6.6 Three-phase voltage source inverter power bridge.

FIGURE 6.7 Simplified single-line diagram of a three-phase inverter.

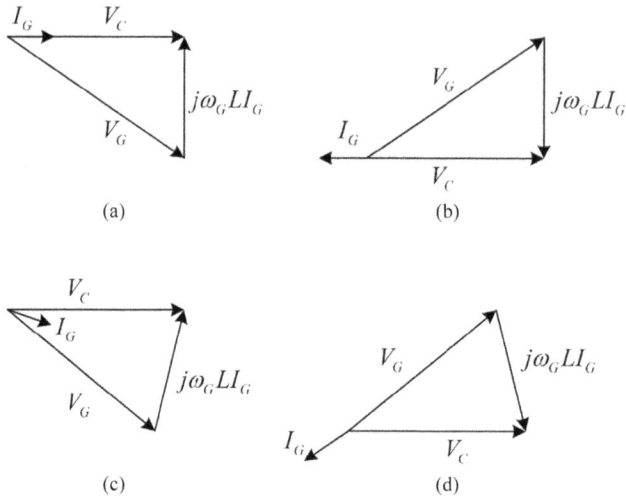

FIGURE 6.8 Phasor diagram according to the operation mode of the inverter: (a) unity power factor rectifying, (b) unity power factor inverting, (c) non-unity power factor rectifying, and (d) non-unity power factor inverting.

output voltage and grid voltage. So that this line impedance limits the current flowing from inverter to grid or vice versa. According to the signal line diagram of the inverter phasor diagram of three-phase voltage source inverter for different operating conditions is given in Figure 6.8.

As shown in Figure 6.8, V_C defines the inverter output voltage while V_G defines grid voltage. Inverter output line current is defined by the voltage and phase difference between inverter output and grid, also the line impedance between inverter output and grid. Since V_C value is controllable with the change of switching state, in order to regulate the line current, control system modifies inverter output voltage. Besides, line impedance is necessary for the control loop to work in a stable way and to control line current. If the filter impedance is too small, excessive current flows can occur, even if there is a minor phase and voltage difference between inverter output and grid.

When the phasor diagrams are examined, it can be seen that in inverting mode inverter output voltage waveform is leading compared to grid voltage waveform. And reactive power flow mostly occurs due to magnitude difference between inverter output voltage and grid voltage. In this phasor diagram, resistance between inverter output and grid is ignored and it can be included, when the filter resistance should be taken into consideration.

6.3 CONTROL STRATEGY OF THREE-PHASE INVERTER

6.3.1 CONTROL METHODS

There are various control methods for three-phase grid connected voltage source inverters. Although the control algorithms for these control methods are different, main purposes are the same. One of these main purposes is stabilizing DC-link voltage at the desired value. For single-stage inverters, which does not contain a DC–DC converter for MPPT function, desired DC-link voltage value of inverter changes to track MPP value. Due to the fact that there is a DC–DC converter to track maximum power point and increase the voltage level, control loop is used to stabilize the DC link at a fixed value by regulating the active power transfer to grid, in order to accomplish the power balance between the power generated from PV strings and power transferred into the grid.

Another main purpose of control method is to maintain the power factor at the desired value. Most of the time inverter is wanted to work at unity power factor, which is the case of zero reactive power. However, in some cases the grid infrastructure can be in the need of reactive power sources or reactive power loads. In those cases, inverter can extract or inject from/to grid in order to support grid. Also, control loop is responsible from harmonic content of line currents injected to grid. There are various control methods and modifications in order to keep total harmonic distortion of line currents at minimum.

The control methods for three-phase grid connected voltage source inverter can be listed as synchronous rotating (dq) reference frame control, stationary ($\alpha\beta$) reference frame control and natural (abc) reference frame control. Synchronously rotated reference frame control, also named as DQ control, is one of the very popular algorithms in three-phase inverter control. Grid angle, θ value is obtained from phase locked loop algorithm applied to line to neutral phase voltages $v_{\text{a-n, b-n, c-n}}$. And by applying park transformation to measured line currents $i_{\text{a, b, c}}$, PI controllable DC quantities of line currents, which are i_d and i_q, are obtained. These two quantities are the main two main controllable variables to control the whole system in a stable way. The transformation phasor diagram of line currents, $i_{\text{a, b, c}}$ to DC quantities of i_d and i_q is given in Figure 6.9.

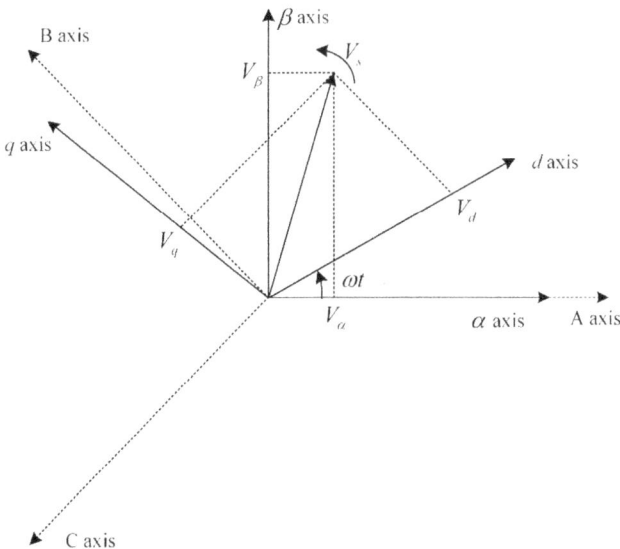

FIGURE 6.9   Synchronous reference frame of phase currents.

The error value between the measured DC link voltage and the reference value of DC link voltage is implemented by PI controller to generate referenced active current component value which is i_d^*. This referenced i_d^* value specifies how much active current is needed in order to keep the DC link voltage at the desired value. After this PI controller, in order to control current, a second inner current control loop takes part. The measured i_d value is subtracted from referenced i_d^* value and goes into a PI controller, in order to generate voltage referenced signal which is v_d^*.

Although in ideal conditions, which is line currents are pure sin wave at fundamental frequency, and unbalance does not exist, i_d and i_q values are pure DC, but this is not the actual case. In real conditions, due to line current unbalance and low order current harmonics, there will be oscillations on i_d and i_q values, and this inner current control loop tries to fixate these oscillations and equalize them to referenced i_d^* and i_q^* value. As the same logic with active current control loop, there is another control loop for reactive power. In order to regulate the reactive power at the desired value, a PI controller named as Q controller generates the referenced i_q^* value, by subtracting measured Q value from referenced Q value. This system can change reactive power according to the grid needs. However, most of the cases, system is desired to work at unity power factor conditions, so that this first PI controller named as Q controller is not needed and i_q^* value is set as zero. Similarly, in the inner id current control loop, in order to generate referenced v_q^* value, there is a PI controller, which is implemented to the subtraction of i_q^* and i_q. This PI controller makes the necessary changes at v_q^* value in order to keep i_q value at zero and regulate the oscillations. Also, v_d and v_q values, achieved from DQ transformation of measured line to neutral voltage values, are feed-forwarded in order to reduce low order harmonics, which are also exist in grid, and also to improve the step response characteristics of three phase inverter at voltage oscillation case at the grid.

The main disadvantage of this control strategy is the poor attenuation of low order current harmonics. Even if, grid voltage and cross-coupling terms are adapted and feed-forwarded, still the compensation capacity is poor. In order to improve that compensation, current compensators, which are specified for fixed low order harmonics especially fifth and seventh, can be adapted. The synchronously rotating reference frame control structure is presented in Figure 6.10.

Another control method is stationary referenced frame control, also known as $\alpha\beta$-control structure, is shown in Figure 6.11. In this control method, instead of transforming line current values to DC quantities of DQ components, line currents are transformed to $\alpha\beta$ values. The PI controllers, used in synchronously rotating reference frame control, does not remove steady state error completely, when the error signal is time-varying. On the other hand, proportional resonant, PR, controllers eliminate the steady state error when the error signal is sinusoidal, which is the case in stationary referenced frame control, due to $\alpha\beta$ values. Because of this reason PR controller has become very popular in control of three-phase grid connected inverters. Although PR controller low order harmonic current compensation capability is good, in order to further improve this capability harmonic compensation, HC, blocks tuned to third, fifth, and seventh harmonics, which are the most dominant low-order harmonics, can be adapted. Besides, PR controller shows good dynamic response characteristics [21].

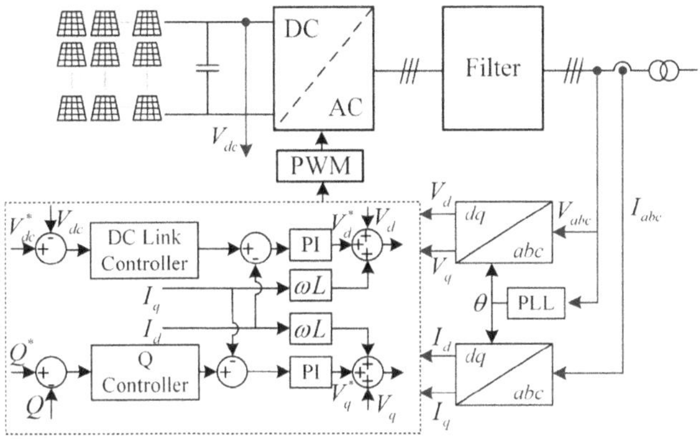

FIGURE 6.10 Synchronous rotating frame control (*dq*-control) structure of three-phase inverter.

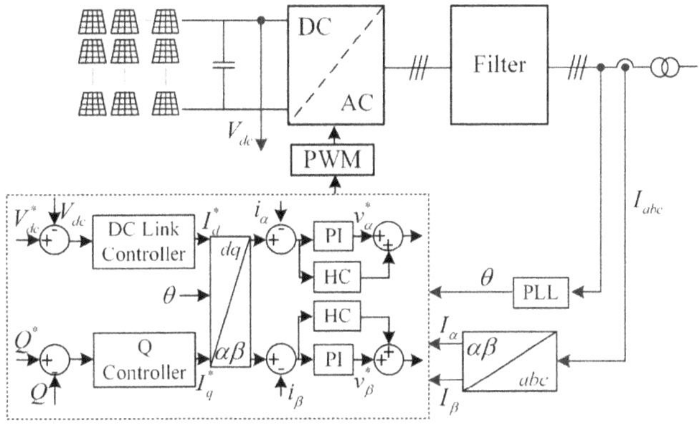

FIGURE 6.11 Stationary frame control (*αβ*-control) structure of three-phase inverter.

According to natural referenced frame control method, each line current is controlled by a different current controller. By implementing PI controllers in DC-link controller and Q controller stage, based on the parameters of value of referenced DC-link voltage subtracted by measured DC link voltage and value of referenced reactive power subtracted by measured reactive power, referenced i_d^* and referenced i_q^* values are obtained. Then these referenced DC quantity current values are transformed to referenced AC quantity line current values i_a^*, i_b^* and i_c^* by implementing inverse DQ transformation. Those referenced current values are subtracted by measured line currents and then this error signal goes into current controller, so that reference PWM signals are obtained. These current controller blocks are achieved by using proportional resonant, PR controller, since the controlled variable is a time varying sinusoidal wave, so that PI controller cannot be used. The structure of natural referenced frame control method also known as *abc* control is given in Figure 6.12.

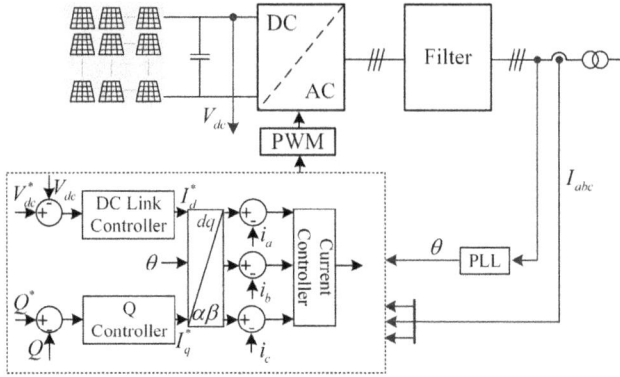

FIGURE 6.12   Natural frame control (*abc*-control) structure of three-phase inverter.

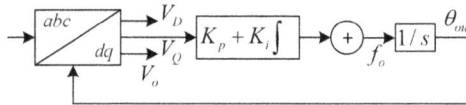

FIGURE 6.13   Synchronous reference frame phase-locked loop block diagram.

The rotational angle, θ, which is needed to implement DQ transformation, is obtained via using phase locked loop block. Despite the fact that rotational angle can be obtained by zero cross detection of line to neutral voltage measurement, due to oscillations, this method can result in wrong calculations of rotational angle. Instead of zero cross detection method, synchronous reference frame method [22] can be used among various PLL algorithms [23]. In this method, line to neutral voltage measurements is transformed via DQ transformation, by using the rotational angle obtained from PLL block. If the rotational angle is correct, the resulting quadrature axis component should be zero. So that, this quadrature axis component of voltage goes into a PI controller to obtain angular velocity. And integral of this angular velocity occurs as the rotational angle, which is the rotational angle used in DQ transformation of the PLL block. When the closed loop control of the PLL block is stabilized, quadrature axis component resulted from DQ transformation occurs as zero, since the rotational angle is set to correct value. Synchronous reference frame phase locked loop block diagram is given in Figure 6.13.

6.4   PULSE WIDTH MODULATION TECHNIQUES

6.4.1   SINUSOIDAL PULSE WIDTH MODULATION

Sinusoidal pulse width modulation is one of the many pulse width modulation methods, which is also simplest and widely used. Basically, in sinusoidal pulse width modulation technique, a reference signal for each phase is generated and compared with a triangle signal, whose frequency is set to switching frequency. Switches are turned on when the reference signal becomes higher than triangular waveform and turned off when the reference signal becomes lower than triangular waveform. MOSFETs in half bridge are in the inverse state with the other MOSFETs in the half bridge, in order to prevent a short-through event by setting an appropriate dead time between turn on instant of a MOSFET and turn off instant of other MOSFET in the half bridge. According to the frequency of triangular waveform, switching frequency of inverter is defined. Each reference waveform has a phase difference of 120° between one and another. According to the frequency of triangular waveform, high frequency switching harmonic occurs at the line currents. According to the turn-on rising time and turn off falling time of semiconductor switch, an appropriate dead time should be used. When the active high complementary switching is used, high-side MOSFET will be turned on, when the reference signal is larger than triangular signal. During turn on and turn off instants, rising edge delay time and falling edge delay times are applied, respectively. Modulation index is defined via division of peak-to-peak magnitude of reference signal by peak-to-peak magnitude of triangular voltage waveform, which is the carrying signal.

During linear modulation in which modulating index, m_a is smaller than 1, inverter line-to-line RMS voltage/DC link voltage is 0.612. As the modulating index increases this value is saturated to 0.78, even if inverter is modulated by square-wave operation. However, when ma is larger than 1, which is the over-modulation case, low order harmonics start to occur in inverter output voltage and hence inverter output current. Because of that, over-modulation is not preferred in SPWM. In order to prevent this, DC link voltage should be selected by considering line to line grid voltage and liner modulation conditions.

6.4.2 Space Vector Pulse Width Modulation

Space vector pulse width modulation, SVPWM is a special pulse width modulation technique that uses a switching sequence of three high side semiconductor device of three phase voltage source inverter, in applications of motor drives and PV inverters. This switching sequence for semiconductor devices establish three sinusoidal currents at the inverter output flowing to the grid. Compared to sinusoidal pulse width modulation technique, SVPWM shows less harmonic distortion in the line currents and inverter output voltages. Also, with SVPWM DC link voltage is used more efficiently, which means less DC link voltage is needed without generating low order harmonics. According to the diagram of three-phase inverter power bridge, which consists of three half bridges, there exist eight possible switching states of high-side power semiconductor switches, since the low-side semiconductor switches are at the inverse state of the high side semiconductor switch states. The space vector projections of these eight switching combinations are called as basic space vectors, and reference voltage vector projection are given in Figure 6.14. Note that the combination of every high side MOSFET switching state on and off conditions, are called as zero vectors and does not generate an output voltage.

As shown in Figure 6.14, reference voltage vector, V^*, is the voltage reference generated by the control loop to transfer desired current to grid. In order to generate

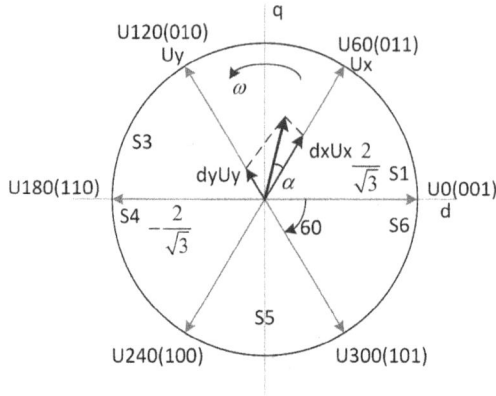

FIGURE 6.14   Reference voltage and basic space vectors projection for SVPWM.

this reference voltage vector, a time average combination of relevant basic vectors that are adjacent to the reference voltage vector at the instantaneous state and zero vectors should be used in a PWM switching period with respective time averages.

6.5   DESIGN OF THREE-PHASE GRID CONNECTED INVERTER

Within the scope of this chapter, a 30 kW two-level three-phase grid-connected SiC inverter has been designed and implemented. The main focus of this design is to transfer solar energy to three-phase 50 Hz 400 V_{l-l} grid in an efficient, robust way and minimize the passive components, in other words input DC capacitor and output LCL filter volume, as far as possible. Another goal is to keep the current harmonics, which are injected to grid, within maximum harmonic current distortion limit according to IEEE STD 519–2014. The implemented system configuration, which can be seen in Figure 6.15, is consisting of multi-string PV system, high-frequency link phase shifted full bridge SiC MPPT converter, which is designed and implemented by Hacettepe University, and three-phase two-level grid connected SiC inverter. While maximum power tracking duty is accomplished by the MPPT converter, VSI inverter is responsible of grid synchronization, grid connection, power transfer and DC-link voltage stabilization. Utilization of high frequency AC link transformer in the MPPT converter enables the use of a conventional two level VSI with 6-pack MOSFET module, since HFlink transformer limits common mode currents, which flows from utilization grid ground to PV modules through earth and PV modules stray capacitance.

During the implementation of multi-string PV system, in order to meet MPPT converter design parameter specifications, CSUN 255–60P PV modules are used as 19 series and five parallel strings configuration, so that PV system open voltage occurs as 712 V and MPPT voltage occurs as 572 V, at the ambient conditions of 20°C ambient temperature, 800 W/m² solar irradiance, according to the electrical characteristics of CSUN 255–60P PV module, which is shown in Figure 6.16.

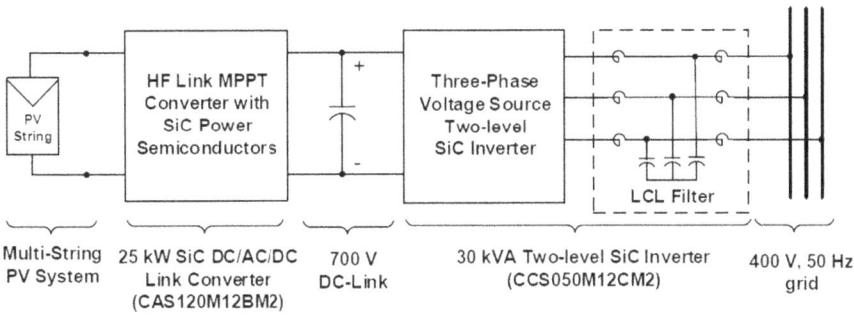

FIGURE 6.15 Experimental setup system configuration of high-frequency link MPPT converter and two-level VSI [24].

Electrical characteristics at Standard Test Conditions (STC)

Module type	CSUN 255-60P	CSUN 250-60P	CSUN 245-60P	CSUN 240-60P	CSUN 235-60P
Maximum Power - Pmpp (W)	255	250	245	240	235
Positive power tolerance	0–3%				
Open Circuit Voltage - Voc (V)	37.5	37.3	37.1	36.9	36.8
Short Circuit Current - Isc (A)	8.88	8.81	8.74	8.67	8.59
Maximum Power Voltage - Vmpp (V)	30.1	29.9	29.7	29.6	29.5
Maximum Power Current - Impp (A)	8.47	8.36	8.25	8.11	7.97
Practical module efficiency	17.46%	17.12%	16.78%	16.44%	16.10%
Module efficiency	15.70%	15.40%	15.09%	14.78%	14.47%

Electrical data refers to standard test conditions (STC): irradiance 1000W/m², AM 1.5 and temperature 25°C the stability uncertainty of power is within ±3%. Certified in accordance with IEC61215/IEC61730/1.2 and IEC 1701.

Electrical Characteristics at Normal Operating Cell Temperature (NOCT)

Module type	CSUN 255-60P	CSUN 250-60P	CSUN 245-60P	CSUN 240-60P	CSUN 235-60P
Maximum Power - Pmpp (W)	188	185	181	178	175
Maximum Power Voltage - Vmpp (V)	28.0	27.9	27.5	27.2	27.0
Maximum Power Current - Impp (A)	6.72	6.64	6.58	6.54	6.48
Open Circuit Voltage - Voc (V)	34.6	34.5	34.2	34.0	33.8
Short Circuit Current - Isc (A)	7.16	7.10	7.02	6.95	6.90

Electrical data refers to normal operating cell temperature (NOCT) irradiance 800W/m², wind speed 1m/s, cell temperature 45°C, ambient temperature 20°C measuring uncertainty of power is within ±3%.

Temperature Characteristics		Maximum Ratings	
Voltage Temperature Coefficient	-0.292%/K	Maximum system voltage(V)	1000
Current Temperature Coefficient	+0.045%/K	Series fuse rating(A)	20
Power Temperature Coefficient	-0.408%/K		

FIGURE 6.16 Electrical characteristics of CSUN 255–60P PV module [25].

6.5.1 Optimum DC Link Voltage

Value of DC link voltage, which is MPPT converter output, and VSI input is kept constant by the inverter in the entire range of operation, no matter the amount of PV power. VSI is supposed to transfer all the power from MPPT converter, into grid and by keeping the value of the DC link constant, this duty is accomplished. As the solar power increases, power output of the MPPT converter increases, and this condition results in the increase of the DC link voltage, if voltage control does not exist. But the control circuitry of the VSI is going to react to keep the DC link voltage constant, by increasing the active power transfer to grid by changing VSI inverter output voltage magnitude and phase with respect to grid voltage magnitude and phase. DC link voltage may vary in some applications, such as single-stage solar voltage source inverter systems in mostly central inverter configurations [26]. In those applications, DC/DC converter stages, which are mostly boost converter serves as MPPT converters, are removed, due to prevent losses occurring in the DC/DC conversion. Maximum power point tracking duty is done by the inverter by varying DC link voltage, such as decreasing or increasing DC link voltage to increase or decrease the current drawn from solar array in a specific manner, to track maximum power point with the various conditions of solar irradiation, ambient temperature and so on. Since in the implemented grid connected PV inverter, MPP tracking is done by the MPPT converter, DC link voltage is constant. DC link voltage, V_{dc} has effects on some features of the inverter, such as efficiency and high frequency current ripple on the filter inductor. For the same output filter inductance value, higher DC link voltage results in increase of the switching frequency current ripple, which increase magnetic losses in the inductor and current harmonic distortion at the switching frequency. Another adverse outcome is the increase of the RMS value of the current through semiconductor switches for the same fundamental output current component, which also results in increase of conduction and switching losses on semiconductor switches. As the V_{dc} increases M, which is the modulation index, decreases when the output voltage value is constant.

In IEC 60038 2002–07 Standard Voltages [27], 50 Hz line-to-line voltage variation ranges are specified. According to this standard, maximum voltage variations in the grid voltages are specified as between +6% and −10%. At the consumer connection points this variation band is increased 4% more and becomes +10% and −14%. These variations in the grid voltage determines variation range of modulation index, M for fixed DC bus voltage under SPWM. Calculated and simulated variation range of M, according to grid voltage disturbances are as shown in Figures 6.17 and 6.18.

According to these plots, which are shown in Figures 6.17 and 6.18, it can be understood that optimum value for DC link voltage is around 700 V. For the supply side voltage upper and lower band limits, inverter does not go into overmodulation, which causes low order harmonics. For the utilization voltage range, inverter works in slightly over-modulated at the upper band limit, which is acceptable. For 650 V

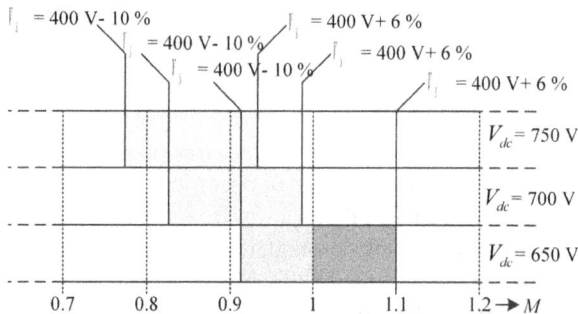

FIGURE 6.17 Range of variation of modulation index M for different fixed values of DC bus voltage and for connected 400 V line-to-line, 50 Hz grid voltage variations upper and lower limits as 400 V +6% −10% as specified in IEC 60038 2002–07 Standard Voltages for the supply side voltage range.

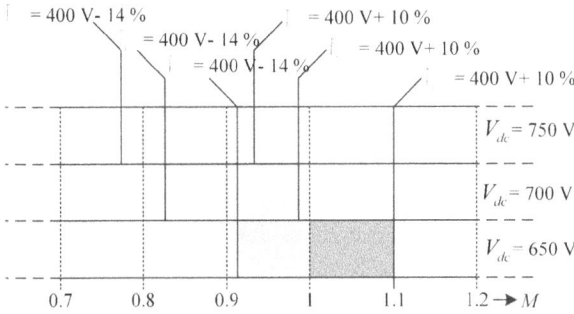

FIGURE 6.18 Range of variation of modulation index M for different fixed values of DC bus voltage and for connected 400 V l-to-l, 50 Hz grid voltage variations upper and lower limits as 400 V +10% −14% as specified in IEC 60038 2002–07 Standard Voltages for the utilization voltage range

DC bus voltage, inverter works highly over-modulated at the upper band limits of both supply side voltage and utilization voltage, which is unacceptable since low order harmonics grow dramatically. In the case of 750 V DC link voltage, inverter does not go into over-modulation in any case for both supply side and utilization side voltage limits, but modulation index M, is too low, which increases losses for the same switching frequency and same output filter. By considering these reasons listed above. DC link voltage is designated as 700 V. Alternative design approaches, such as third harmonic injection [28,29], enable to reduce DC link voltage to nearly 600 V without causing low-order harmonics. So, third harmonic injection method could be used to reduce switching and conduction losses by reducing DC link voltage in future works.

6.5.2 OPTIMUM SWITCHING FREQUENCY

For many modulation techniques, such as SPWM and SVPWM, as the switching frequency increases, frequency of high order harmonics, which are also known as side-band harmonics occurring both sides of switching frequency and its' multiplies, also increases [30]. So that, output filter corner frequency, in other words resonant frequency of LCL filter, can be set to a higher value in order to reduce the volume of passive components. However, increasing switching frequency has a drawback in terms of losses. As the switching frequency increases, switching loss occurring in power semiconductors proportionally increases with the frequency. Traditional Si based power semiconductors, such as IGBT, has significantly larger turn-on, turnoff times with respect to SiC based power semiconductor. So that, with the emergence of SiC power MOSFETs, switching frequency can be increased more, in comparison to IGBT switching frequency, for the same power rating of the inverter and for the same switching loss. But, benefits of increasing switching frequency, in terms of lowering the size of passive components, start to saturate at one point in the viewpoint of control stability. Reducing inductor component of output filter causes stability problems since voltage and phase difference between inverter output voltage and grid voltage is necessary to control power flow to grid. Output filter inductor is the component which compose the phase and voltage difference, and current flow occurs according to this difference between inverter output and grid voltage according to impedance between them, as can be understood from single line diagram of three phase inverter. Reducing impedance too much, between inverter output and grid voltage causes massive current flows, even with a slight voltage difference. This condition is also a problem, in terms of low order harmonics which exist in grid voltage. Low-order voltage harmonics in the grid cause low order current flows, despite control circuitry current loop attenuation. Because of these reasons, output filter inductance has a minimum value which has shown in output filter design section, so that increasing switching frequency too much has no advantage after one point. In order to find optimum switching frequency, loss calculations have been done for SiC MOSFETs conduction loss, switching loss, SiC body diode conduction loss, output LCL filter copper loss and LCL filter core loss. These calculations are made for different switching frequencies such as 10, 20, and 30 kHz, which are seen to be feasible for inverter power rating and output LCL filter specifications. Loss analysis has been verified analytically, in addition to manufacturer loss analysis tool [31]. Detailed analytical loss calculation of power semiconductor losses and LCL output filter loss are also investigated in power loss calculation section. Loss components, such as the wiring from power source to inverter, inverter to grid and between discrete components of the inverter are ignored in the calculations. Also, only fundamental frequency current component, which is 50 Hz, has been considered during calculation of losses in SiC MOSFETs, even if there are switching current ripples about 25% of fundamental current superimposed on main current component, due to ease calculation. So, in fact, the real semiconductor and wiring loss before filtering is slightly higher than calculated value. When calculating the loss of LCL output filter losses, LCL filter is modified for the switching frequency, so that switching current ripple is the 25% of the fundamental 50 Hz current, due to consistency of switching frequency selection. Total power loss and distribution of loss components can be seen in Figure 6.19. These calculations are made for output power of the inverter, $P_o = 22.3$ kW which is the maximum achievable power from the installed PV system.

According to these results, switching frequency has been selected as 20 kHz, due to following reasons. 10 kHz switching frequency has almost no advantage in terms of losses and LCL filter size needs to be increased significantly. LCL filter copper loss increases dramatically, in order to have the same switching current ripple. In the case of 30 kHz switching frequency, LCL filter volume reduction according to 20 kHz switching frequency is not very remarkable, in addition, switching loss of SiC MOSFETs starts to be a considerable amount. Also, because of mentioned control loop stability problems, due to insufficient impedance between inverter output and grid voltage, 30 kHz switching frequency is not considered as advantageous. So that switching frequency is selected as 20 kHz.

6.5.3 Output Filter Design

In order to attenuate high-order harmonics, which resulting from switching at sidebands of the switching frequency, a low pass filter, mostly L or LCL type, is placed

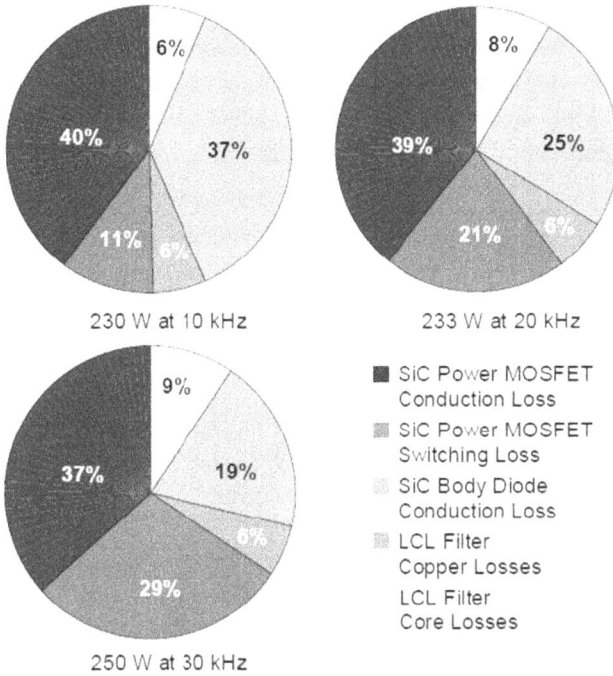

FIGURE 6.19   Total loss and distribution of loss components.

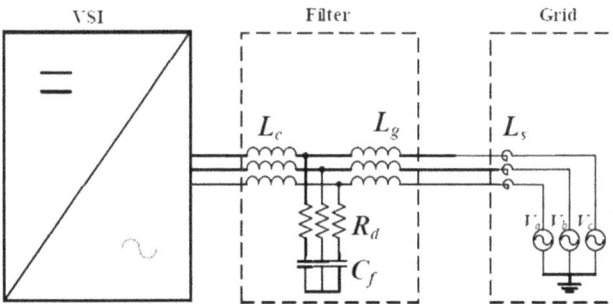

FIGURE 6.20   LCL filter-based filtering circuit diagram of three-phase VSI.

between output of the VSI and grid. LCL filter harmonic attenuation is considerably larger than L filter, due to the fact that LCL filter is a second order low pass filter, while L filter is a first order filter. Because of this reason, LCL filter needed inductance value is much lower compared to L filter, for the same harmonic attenuation, and this condition enables volume reduction of filter, lower filter loss and better current filtering. As can be seen in Figure 6.20, LCL filter is composed of four components, L_c as inverter side inductor, L_g as grid side inductor, C_f as filtering capacitor and R_d as damping resistor. Although for this design damping resistor, R_d is placed in series with the filtering capacitor C_f, there are several other methods for damping resistor placement, such as placing ramping resistor R_d in parallel with filtering capacitor, in series with inverter side inductor L_c or in parallel with inverter side inductor. All of this design type has some advantages and disadvantages.

As design methodology of LCL filter, there are several design methods [32,33], and for all of them, the main design purpose is to maximize the high order harmonic filtration, caused from switching, with minimum reactive power consumption and filter volume. According to step-by-step design methodology [34] of LCL filter, LCL filter component values have been calculated in the following procedure.

Furthermore, resonance phenomenon should be taken into consideration carefully when using LCL filter. For an LCL filter, at the resonance frequency harmonics, whose frequency is close to resonance frequency, are highly boosted. So that, resonance phenomenon causes very high harmonic distortion in line currents, stability issues even loss of synchronization of inverter with grid, if no precautions are taken. Firstly, resonance frequency should be selected according that resonance frequency should not be close to harmonic distortion sources in order to avoid boosting harmonics.

In order to damp the boosting effect at the resonance frequency, active damping or passive damping should be utilized [35,36]. In this research work, passive damping is used by utilizing series damping resistor, R_d in series with filtering capacitor. R_d value affects the loss value on damping resistor, damping factor at resonance frequency and attenuation of switching frequency, so that resistance value must be carefully selected. Apart from calculations of LCL filter components values, some modifications should be made with these components, in order to reduce volume of the filter, in the manner of providing enough attenuation at the switching frequency. By optimizing filter parameters, resulting components values are found as $L_c = 250\ \mu H$, $C_f = 15\ \mu F$, $L_g = 50\ \mu H$, and $R_d = 0.22\ \Omega$. For these values, resonance frequency viewed from inverter side, is found as 6.66 kHz, which is $f_{sw}/3$, and provides enough

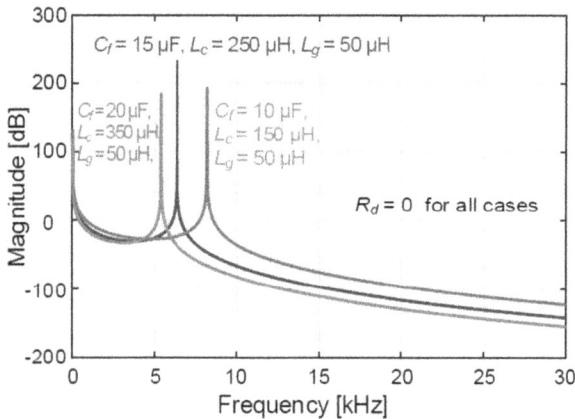

FIGURE 6.21 From inverter to grid transfer characteristics of various sets of components values of LCL filters.

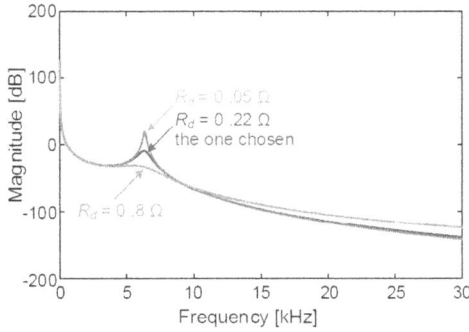

FIGURE 6.22 From inverter to grid transfer characteristics of selected LCL filter for various damping resistor values.

attenuation of switching harmonics. For various set of filter components, without damping resistor, transfer characteristics of LCL filters is given in Figure 6.21.

As can be seen in Figure 6.21, all LCL filter sets provide enough attenuation at switching frequency about – 100 dB. Red colored filter ($L_c = 350\,\mu$H, $C_f = 20\,\mu$F, $L_g = 50\,\mu$H) is not chosen, since converter side inductor, L_c increases volume of filter almost 40%, and that much attenuation is not needed. For the green colored filter ($L_c = 150\,\mu$H, $C_f = 10\,\mu$F, $L_g = 50\,\mu$H), although it provides enough attenuation at switching frequency, and it has the minimum LCL filter volume, it is not chosen either, because of stability issues resulting from narrow control range of voltage and phase shift angle, causing output power oscillations and instability, since L_c is too low. Field work experience has shown that, larger L_c value enables to increase proportional and integral term of PI current control and reduce of low order harmonics through current control. Boosting effect of harmonics, at resonant frequency, also can be seen in Figure 6.21. In order to reduce and attenuate magnitude of amplification under 0 dB for several values of damping resistors Rd, change of transfer characteristics plotted, as can be seen in Figure 6.22, for the selected LCL filter ($L_c = 250\,\mu$H, $C_f = 15\,\mu$F, $L_g = 50\,\mu$H).

For the colored line which R_d is 0.8 Ω, filter resonance frequency amplification is properly damped, but filter attenuation value is reduced because of excessive damping of filter. Because of this, further increase of damping resistor value makes switching harmonics significant. Also, loss of damping resistor is relatively high for this value. In the case of selecting damping resistor value as 0.05 Ω, transfer characteristics of LCL filter is as green colored line. Attenuation of high order harmonics is almost the same with the un-damped filter, but at resonance frequency, damping of amplification is not enough. For $R_d = 0.22$ Ω, damping of amplification at resonance frequency is enough, since magnitude is under 0 dB, also attenuation at the switching frequency does not change significantly. So that value of damping resistor, R_d selected as 0.22Ω.

In order to compare designed LCL filter transfer characteristics, in terms of inverter to grid and grid to inverter, for both transfer functions, LCL filter transfer characteristic is plotted as is can be seen in Figure 6.23.

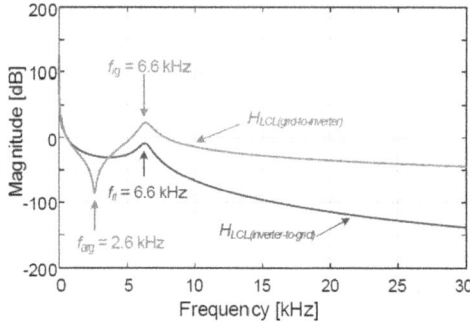

FIGURE 6.23 From inverter to grid (blue) and from the grid to inverter (red) transfer characteristics of designed LCL.

As shown in Figure 6.23, for grid to inverter transfer characteristic of LCL filter, line forms a dip and a peak as stated earlier and resonance frequency of boosting is the same resonance frequency with inverter to grid transfer function resonance frequency.

6.6 SUMMARY

In this chapter a grid connected voltage source three-phase inverter with SiC MOSFET module has been designed and implemented in order to work with a phase-shifted full bridge MPPT converter, in such a way that these two converters compose a full system solution PV supply to transfer PV energy into the grid in an efficient and robust way. Emergence of commercial SiC based power MOSFETs, which have low turn-on and turn-off times, has enabled to increase switching frequency compared to traditional Si based switching components. This circumstance was one the main motives of this chapter in the scopes of increasing the switching frequency in order to reduce passive component volumes of inverter, such as LCL filter and DC link capacitor volume and decreasing losses which leads to reduce in heat sink volume and having better efficiency.

In the design stage of three-phase inverter, inverter parameters such as passive component values, rated voltage selection, pulse with modulating method and inverter control method have been selected, and inverter designed based on various design methodologies.

Optimum switching frequency has been selected by making the trade-off between decrease of passive components' volume and increase of losses occurring in the power stage of inverter with the increase of switching frequency into consideration. Besides, as explained in the design section, filter size was not completely depended on switching frequency. In order to maintain the stability of inverter control loop, a moderate inductance value between inverter output and grid was needed. So that, after some point, increase of switching frequency was not advantageous in any ways.

LCL filter parameters have been determined to have enough attenuation at the switching frequency primarily. Although smaller size of LCL filter had enough attenuation at the switching frequency, in order to be in the safe side by considering control loop stability of inverter, LCL filter parameters determined accordingly, as stated in the design section. Active damping method was not preferred in LCL filter since the losses occurring due to the passive damping method was tolerable. SPWM was used as the modulating method due to the fact that its simplicity and to make loss calculations more accurate in order to show SiC MOSFETs' advantages in a more explicit way, despite the fact that SVPWM modulating method was more advantageous via enabling the reduction of necessary DC link voltage, so that decreasing losses occurring in SiC MOSFETs.

Synchronously rotating reference frame control was used as the control method of the three-phase inverter, again for its simplicity. Although implementation of this control method was easy, it was inefficient in terms of attenuating low order harmonics resulting from grid voltage harmonics, mainly fifth and seventh orders. In order to reduce the harmonics, grid line-to-neutral voltages had been imitated at the output of the inverter, apart from the necessary changes that are needed to control active and reactive power flow.

Loss components, such as switching and conduction losses of SiC MOSFETs and LCL filter losses, were calculated in the design section for various output powers in order to determine heat sink requirement and to forecast efficiency of three-phase inverter. According to efficiency calculations inverter efficiency at rated output power arose as 99%. With regard to calculated loss components, heat transfer circuitry of power stage had been deduced and necessary heat sink requirement was calculated. Heat sink selection was made based on this value.

6.7 QUESTION

1. What is PV Module?
2. What is surge Arrestor?
3. Explain Photo voltaic PV model and characteristics?
4. What is the Equivalent circuit of multi-string PV system?
5. What is maximum power point tracker (MPPT)?
6. What is mission profile of solar irradiance?
7. What is transformer less inverter? How it is different from inverter with transformer?
8. Why reliability of PV-based system is important?
9. What are the different configuration of PV systems?
10. What is SiC Mosfet? How it is different from other MOSFETs?
11. Explain Radar Chart comparison of Si- and SiC-based devices?
12. What is bandgap devices of SiC- and GaN-based semiconductors?
13. Draw Simplified single line diagram of three-phase inverter and explain its working?
14. What are different control scheme used for three phase grid-connected inverters?
15. What is synchronous rotating phase control? Explain.
16. What is PR-based control for solar inverters?
17. What is the natural frame control (*abc*-control) structure of a three-phase inverter?

18. What is a Phase Lock loop? How does it work?
19. What is pulse width Sinusoidal modulation?
20. is space vector Pulse Width Modulation?
21. What is the method to decide on output Filter design?

REFERENCES

[1] S. R. Dzonde Naoussi, S. Kenfack Tsobzé, and G. Kenne, "A new approach for islanding management in PV system using frequency modulation technique and simple adaptive load shedding algorithm," *SN Appl. Sci.*, vol. 1, no. 12, p. 1642, Dec. 2019, doi: 10.1007/s42452-019-1690–y.

[2] D. Kolantla, S. Mikkili, S. R. Pendem, and A. A. Desai, "Critical review on various inverter topologies for PV system architectures," *IET Renew. Power Gener.*, vol. 14, no. 17, pp. 3418–3438, Dec. 2020, doi: 10.1049/iet-rpg.2020.0317.

[3] ABB, "ABB String Inverters," Product catalogue, 2014.

[4] J. Rabkowski, D. Peftitsis, and H.-P. Nee, "Design steps towards a 40-kVA SiC inverter with an efficiency exceeding 99.5%," in *2012 Twenty-Seventh Annual IEEE Applied Power Electronics Conference and Exposition (APEC)*, Feb. 2012, pp. 1536–1543. doi: 10.1109/APEC.2012.6166024.

[5] B. Pakkiraiah and G. D. Sukumar, "Research survey on various MPPT performance issues to improve the solar PV system efficiency," *J. Sol. Energy*, vol. 2016, pp. 1–20, Jul. 2016, doi: 10.1155/2016/8012432.

[6] "IEEE Guide for Terrestrial Photovoltaic Power System Safety," *IEEE Std 1374-1998*, pp. 1–64, 1998, doi: 10.1109/IEEESTD.1998.88280.

[7] A. Z. AL Shaqsi, K. Sopian, and A. Al-Hinai, "Review of energy storage services, applications, limitations, and benefits," *Energy Reports*, vol. 6, pp. 288–306, Dec. 2020, doi: 10.1016/j.egyr.2020.07.028.

[8] J. Ahmad, A. Ciocia, S. Fichera, A. F. Murtaza, and F. Spertino, "Detection of typical defects in silicon photovoltaic modules and application for plants with distributed MPPT configuration," *Energies*, vol. 12, no. 23, p. 4547, Nov. 2019, doi: 10.3390/en12234547.

[9] L. Idoko, O. Anaya-Lara, and A. McDonald, "Enhancing PV modules efficiency and power output using multi-concept cooling technique," *Energy Reports*, vol. 4, pp. 357–369, Nov. 2018, doi: 10.1016/j.egyr.2018.05.004.

[10] A. Razak, Y. Irwan, W. Z. Leow, M. Irwanto, I. Safwati, and M. Zhafarina, "Investigation of the effect temperature on photovoltaic (PV) panel output performance," *Int. J. Adv. Sci. Eng. Inf. Technol.*, vol. 6, no. 5, p. 682, Oct. 2016, doi: 10.18517/ijaseit.6.5.938.

[11] M. M. Rahman, M. Hasanuzzaman, and N. A. Rahim, "Effects of various parameters on PV-module power and efficiency," *Energy Convers. Manag.*, vol. 103, pp. 348–358, Oct. 2015, doi: 10.1016/j.enconman.2015.06.067.

[12] F. Iov, M. Ciobotaru, D. Sera, R. Teodorescu, and F. Blaabjerg, "Power electronics and control of renewable energy systems," in *2007 7th International Conference on Power Electronics and Drive Systems*, Nov. 2007, p. P-6-P-28. doi: 10.1109/PEDS.2007.4487668.

[13] M. Ciobotaru, R. Teodorescu, and F. Blaabjerg, "Control of single-stage single-phase PV inverter," in *2005 European Conference on Power Electronics and Applications*, 2005, pp. 10 pp.-P.10. doi: 10.1109/EPE.2005.219501.

[14] Institute of Electrical & Electronics Engineers, "*IEEE 929–2000 Recommended Practice for Utility Interface of Photovoltaic (PV) Systems*," 2000.

[15] D. Meneses, F. Blaabjerg, Ó. García, and J. A. Cobos, "Review and comparison of step-up transformerless topologies for photovoltaic AC-module application," *IEEE Trans. Power Electron.*, vol. 28, no. 6, pp. 2649–2663, Jun. 2013, doi: 10.1109/TPEL.2012.2227820.

[16] M. Liserre, R. Teodorescu, and P. Rodriguez, *Grid Converters for Photovoltaic and Wind Power Systems*. Wiley, 2011.

[17] F. Blaabjerg, Y. Yang, D. Yang, and X. Wang, "Distributed power-generation systems and protection," *Proc. IEEE*, vol. 105, no. 7, pp. 1311–1331, 2017, doi: 10.1109/ JPROC.2017.2696878.

[18] S. B. Kjaer, J. K. Pedersen, and F. Blaabjerg, "A review of single-phase grid-connected inverters for photovoltaic modules," *IEEE Trans. Ind. Appl.*, vol. 41, no. 5, pp. 1292–1306, Sep. 2005, doi: 10.1109/TIA.2005.853371.

[19] J. Biela, M. Schweizer, S. Waffler, and J. W. Kolar, "SiC versus Si—evaluation of potentials for performance improvement of inverter and DC–DC converter systems by SiC power semiconductors," *IEEE Trans. Ind. Electron.*, vol. 58, no. 7, pp. 2872–2882, Jul. 2011, doi: 10.1109/TIE.2010.2072896.

[20] C. Sintamarean, F. Blaabjerg, H. Wang, and Y. Yang, "Real field mission profile oriented design of a SiC-based PV-inverter application," in *2013 IEEE Energy Conversion Congress and Exposition,* Sep. 2013, pp. 940–947. doi: 10.1109/ECCE.2013.6646804.

[21] A. G. Yepes, F. D. Freijedo, O. Lopez, and J. Doval-Gandoy, "Analysis and design of resonant current controllers for voltage-source converters by means of Nyquist diagrams and sensitivity function," *IEEE Trans. Ind. Electron.*, vol. 58, no. 11, pp. 5231–5250, Nov. 2011, doi: 10.1109/TIE.2011.2126535.

[22] V. Kaura and V. Blasko, "Operation of a phase locked loop system under distorted utility conditions," in *Proceedings of Applied Power Electronics Conference. APEC '96*, vol. 2, pp. 703–708. doi: 10.1109/APEC.1996.500517.

[23] L. R. Limongi, R. Bojoi, C. Pica, F. Profumo, and A. Tenconi, "Analysis and comparison of phase locked loop techniques for grid utility applications," in *2007 Power Conversion Conference - Nagoya*, Apr. 2007, pp. 674–681. doi: 10.1109/PCCON.2007.373038.

[24] S. Öztürk, M. Canver, I. Çadırcı, and M. Ermiş, "All SiC grid-connected PV supply with HF link MPPT converter: System design methodology and development of a 20kHz, 25 kVA prototype," *Electronics*, vol. 7, no. 6, p. 85, May 2018, doi: 10.3390/electronics7060085.

[25] SecondSol, "CSUN255–60p Poly PowerGuard-Versicherung Speciality Insurance Services," 2015.

[26] K. Fujii, Y. Noto, M. Oshima, and Y. Okuma, "1-MW solar power inverter with boost converter using all SiC power module," in *2015 17th European Conference on Power Electronics and Applications (EPE'15 ECCE-Europe)*, Sep. 2015, pp. 1–10. doi: 10.1109/EPE.2015.7309080.

[27] IEC, "*IEC 60038, IEC Standard Voltages*," 2009.

[28] J. Feng, H. Wang, J. Xu, M. Su, W. Gui, and X. Li, "A three-phase grid-connected microinverter for AC photovoltaic module applications," *IEEE Trans. Power Electron.*, vol. 33, no. 9, pp. 7721–7732, Sep. 2018, doi: 10.1109/TPEL.2017.2773648.

[29] J. W. Kimball and M. Zawodniok, "Reducing common-mode voltage in three-phase sine-triangle PWM with interleaved carriers," in *2010 Twenty-Fifth Annual IEEE Applied Power Electronics Conference and Exposition (APEC)*, Feb. 2010, pp. 1508–1513. doi: 10.1109/APEC.2010.5433431.

[30] Y.-M. Chen, K.-Y. Lo, and Y.-R. Chang, "Multi-string single-stage grid-connected inverter for PV system," in *2011 IEEE Energy Conversion Congress and Exposition*, Sep. 2011, pp. 2751–2756. doi: 10.1109/ECCE.2011.6064138.

[31] "SiC and GaN Solutions SpeedFit Design Simulator | Wolfspeed." https://www.wolfspeed.com/tools-and-support/power/speedfit/ (accessed Jul. 30, 2022).

[32] K. Jalili, "*Investigation of Control Concepts for High-Speed Induction Machine Drives and Grid Side Pulse-Width Modulation Voltage Source Converters*," Thesis, 2009.

[33] T. Sobh, K. Elleithy, A. Mahmood, and M. Karim, Eds., *Innovative Algorithms and Techniques in Automation, Industrial Electronics and Telecommunications*. Dordrecht: Springer Netherlands, 2007. doi: 10.1007/978-1-4020-6266-7.

[34] M. Liserre *, F. Blaabjerg, and A. Dell'Aquila, "Step-by-step design procedure for a grid-connected three-phase PWM voltage source converter," *Int. J. Electron.*, vol. 91, no. 8, pp. 445–460, Aug. 2004, doi: 10.1080/00207210412331306186.

[35] M. Liserre, A. Dell'Aquila, and F. Blaabjerg, "Genetic algorithm based design of the active damping for a LCL-filter three-phase active rectifier," in *Eighteenth Annual IEEE Applied Power Electronics Conference and Exposition, 2003. APEC '03.*, vol. 1, pp. 234–240. doi: 10.1109/APEC.2003.1179221.

[36] M. Liserre, A. Dell'Aquila, and F. Blaabjerg, "Stability improvements of an LCL-filter based three-phase active rectifier," in *2002 IEEE 33rd Annual IEEE Power Electronics Specialists Conference. Proceedings (Cat. No.02CH37289)*, pp. 1195–1201. doi: 10.1109/PSEC.2002.1022338.

7 Advance Control Feature for Grid Connected PV System

7.1 INTRODUCTION

The advancement in power electronic converters has also resulted in the modification of the controllers over time. With more and more distributed generations (DGs) interacting with each other, it is required that the controllers of all the DGs can interact and operate consistently. To further enable the smooth and safe operation of DGs, distribution companies and governments collaborate to form a grid code that ensures the actions to be taken during a faulty condition. Most of the grid codes specify that once the fault is identified, then the DGs need to disconnect from the grid for safe operation. During the fault, DGs should supply power to the local load, and as the utilities are back online, the DGs must reconnect. However, the implementation of grid codes presents a lot of challenges ranging from fast fault detection to smooth disconnection and reconnection of the DGs to utilities. DGs also provide an opportunity to support the utilities by their capability to inject reactive power into the grid and try to stabilize the system. All these stages of operation from fault detection to reactive power injection and smooth disconnection to resynchronization of DGs before reconnection should be done by the controller and the operation needs to do efficiently so that the system does not present a cyclic behavior of disconnection due to the unintentional islanding and false fault detection. In this chapter, an overview related to islanding detection and a brief about multiple islanding detection techniques is presented. The ride-through operation parameters are also discussed in the chapter, and few methods to achieve the ride-through are also discussed. In case the system is not recovered by the ride-through application, an account related to disconnection and reconnection of the DGs from the grid is also presented, discussing the different parameters which need to be considered during such operation.

7.2 ISLANDING DETECTION

The islanding detection (ID) mechanism functions as a safety procedure in the control of grid-connected inverters, identifying anomalous grid behavior based on grid codes. Furthermore, the ID mechanism must detach the DGs from the grid to function with local loads, according to the parameters provided with the grid codes. This ID-based detection and disconnection method must be completed within a certain amount of time, as required by various grid standards. As a result, it is critical to spot

DOI: 10.1201/9781003257189-7

irregularity quickly and properly. During this procedure, it was discovered that the DGs encounter the following challenges:

- If there is a mismatch between power demand and supply, the performance of DGs might be unsteady.
- The maintenance crew operating on the line may face a life-threatening situation because of unmanaged islanding.
- Once the utility problem has been resolved, the DGs must be reconnected to the grid with utmost caution and safety regulations.

Multiple ID methods have been created in the literature throughout the years, considering varied grid standards and constraints. The ID methods can be classified as active, passive, remote, hybrid, or data-driven, depending on the location and parameters employed for the operation. The next sections provide a full review of all the ID schemes.

7.2.1 PASSIVE ISLANDING DETECTION METHOD

Because the multiple operational parameters are evaluated at the point of common coupling (PCC) to identify the irregularities in the system, passive ID is one of the most cost-effective ID techniques. For ID, this approach employs a variety of sensors to determine the threshold between the operational and non-functioning zones. The DGs are isolated from the utility using the relay switches that are easily available in the power system for protection. The existence of non-detection zone (NDZ) is one of the key limitations of passive ID. Regardless of the disadvantages, it is one of the most extensively used methods for identification.

Table 7.1 lists the most often used passive islanding detection methods (IDMs), and Figure 7.1 depicts a simplified structure for integrating passive IDM with a photovoltaic (PV) system.

7.2.1.1 Change in Power Rate

This IDM monitors the system's power output to estimate its derivative and keep it below a certain threshold. The variance in power during normal operation is low; however, in the event of a malfunction, the measured power may indicate a substantial variation. The threshold value must be carefully determined since it must distinguish the load-switching event from the islanding event [1,2]. To identify major power mismatches, the voltage index of numerous DGs is monitored in reference [4]. In addition, the forced Helmholtz oscillation is utilized to detect power discrepancy in reference [3]. The fundamental advantage of this technique over other oscillators is the chaotic and normal movements. However, because load-switching situations may be misclassified, the threshold selection might render the system unstable.

7.2.1.2 Rate of Change of Frequency (ROCOF)

When the value of df/dt exceeds a predetermined threshold, this frequency-based IDM generates a trip signal. The threshold is typically between 0.1 and 1.2 Hz/sec [17] as per the industry norms. To avoid tripping, the difference between a start

TABLE 7.1
A Comparison of the Various Passive ID Approaches

Method	Merit	Demerit
Rate of change of power [1–4]	Effective method when there is a case related to a large power mismatch.	The system can be deemed unreliable because of the selection of threshold values.
Rate of change of frequency [5, 6]	The response time is faster compared to the other detection techniques.	Not very ideal for the medium or large-scale DGs as there is a threshold selection involved.
Over-/under-voltage and frequency [7,8]	Low cost of implementation makes it one of the most desirable methods.	The large NDZ may impact the response time of the system.
Change of impedance [9,10]	It is derived from linked topology and may be used to extract features.	This method's shortcoming is the annoyance of trip initialization.
Phase jump [11]	It can be implemented easily by modifying PLL in the controller.	Issue with ID if the local demand is being met by the DGs.
Harmonic distortion [12]	The NDZ is low compared to the conventional method used for ID.	The high Q value can also be a concern while performing the operation.
Voltage unbalance [13,14]	Highly effective in recognizing unbalances for three-phase system operation.	A system is incompatible with a single-phase system.
Rate of change of frequency over power [15,16]	It can detect a minor power difference between the DG and the load.	Setting up a cut-off range with the wrong threshold might result in erroneous detection.

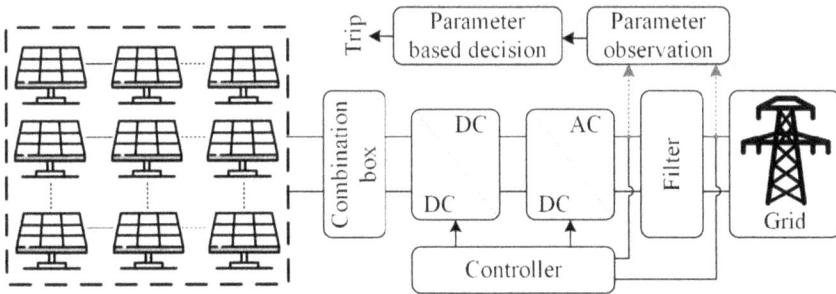

FIGURE 7.1 An overview of the passive ID approach using a PV system.

time-frequency and a run time-frequency must be established. The following equations can also be used to construct a relationship between power and frequency:

$$P + jQ = \frac{V_{a1}V_{b1}\angle\theta}{Z_1} \qquad (7.1)$$

$$P = \frac{V_{a1}V_{b1}}{Z_1}.\mathrm{Sin}(\theta) \qquad (7.2)$$

$$Q = \frac{V_{a1}}{Z_1}\left(V_{b1}.\text{Cos}(\theta) - V_{a1}\right) \qquad (7.3)$$

The active and reactive power is denoted by P and Q, respectively. V_{a1} and V_{b1} are the voltages at two points, and Z_1 represents impedance.

By monitoring q-$axis$ current controller models, a frequency deviation-based ID in reference [5]. The frequency deviates from the reference current value in the event of a failure. Similarly, to identify the islanding situation in reference [6], the frequency variation is observed simultaneously with the phase deviation.

7.2.1.3 Over-/Under-Voltage and Frequency

In the literature, the over/under-voltage and frequency approaches are the most often used passive IDMs. A power discrepancy between generation and consumption will cause a frequency shift as well as a reactive power mismatch [18]. The frequency shift is checked to see whether it is less than 3%, but the reactive power mismatch causes voltage fluctuation, which causes the trip signal to be created if the voltage changes above the pre-set values and is unable to recover. When compared to frequency-based IDMs, most studies show that volage-based IDMs have higher accuracy and greater performance [19].

When compared to the separate execution of each aspect in the technique, Liu et al. [8] propose a combination of both over/under frequency and over/under-voltage ways to obtain reduced NDZ. The method is more reliable and has a lower rate of misinterpretation. In addition, Reddy and Sreeraj [7] develop a feedback-based IDM for monitoring by merging it with once-cycle feedback control.

7.2.1.4 Change in Impedance

The DGs have a low impedance in normal operation, but when connected to the power grid, the impedance is larger than the source. If an imbalanced load causes islanding, PCC's impedance monitoring allows IDM to operate, and as soon as an impedance change is detected, the DG is detached. For feature extraction and threshold selection for ID, a simple method is utilized in reference [10]. In addition, the method in reference [9] seeks to distinguish between islanding and system distribution handling capabilities owing to capacitor switching, load switching, and load imbalance.

7.2.1.5 Phase Jump

The phase angle difference between the voltage waveform and the reference waveform is evaluated in this ID method. In most cases, the phase shift happens to collect the system's power shift [20]. The relay will disconnect the DGs in 300 ms if the mismatch is more than 33% [21]. The fluctuation in voltage phase angle is regarded as feedback in reference [11]. As a result, the phase angle may produce a change in active and reactive power during the post-islanding situation. This variance causes a false trip situation by causing variations in frequency and voltage.

7.2.1.6   Harmonic Distortion

The total harmonic distortion (THD) value is specified as a threshold based on grid code criteria [22], and the relay disconnects the DGs whenever the THD exceeds the threshold. The grid codes set the voltage and current signal thresholds at 5% and 10%, respectively [23]. A technique for ID is proposed in reference [12] by monitoring voltage imbalance and THD. This method has a flaw in that it has a high Q factor, which leads to incorrect ID and disconnection.

7.2.1.7   Voltage Unbalance

The voltage imbalance in the grid-connected operation of the DG system was observed using this passive IDM, which revealed the change in reactive power. Overvoltage occurs when the reactive power in the system is too high, whereas under-voltage occurs when the reactive power is too low. The relay will disconnect the DGs from the grid if the voltage falls below a certain threshold value [19]. In reference [14], Laaksonen discusses a strategy that combines the voltage imbalance method with harmonic variation. In reference [13] a component-based model that considers the voltage signal is created. The new technology is speedier and has the capability of detecting anomalies even under perfectly balanced power situations. The technique's shortcoming is that it is unlikely to be implemented in single-phase systems.

7.2.1.8   Rate of Change of Frequency over Power

When compared to the ROCOF technique, this IDM has a smaller NDZ. Small and medium-sized power systems are the most common targets for this strategy [15]. A pre-set threshold value is used to perform the identification. The method may also be used for zero power ID with a modest or non-existent NDZ [14]. Furthermore, the threshold-based configuration has been discovered as a source of false ID and disconnection.

7.2.2   ACTIVE ISLANDING DETECTION METHODS

A little disturbance is injected at the PCC during the active IDM. Because of its modest size, this disturbance does not produce any transients in the system's operation. When compared to passive detection approaches, the NDZ present in active IDM is substantially less. An active IDM has a drawback: it causes grid instability. To identify the various stages of operation, the variation in the injected perturbation is noted. Table 7.2 lists the most often used active IDMs in the literature, and Figure 7.2 shows a quick summary of the arrangement.

7.2.2.1   Active and Reactive Power Injection

To minimize the NDZ and precisely identify islanding, the active and reactive power injection approach injects a pulse current. The disturbance signal is injected into the grid via the voltage source converter $d-axis$ or $q-axis$ current controllers in reference [45]. The signal injection changes the amplitude of the voltage at PCC in the case of a $d-axis$-based controller, whereas a frequency deviation occurs at PCC in the case of a $q-axis$-based controller during the islanding condition. A negative

TABLE 7.2

A Comparison of Active ID Algorithms

Method	Merit	Demerit
Active and reactive power injection [24–29]	The power injection influences the rotating frame of reference, which improves detection accuracy.	Because of the voltage spike at the distribution end, system-related concerns might be a constraint.
Active frequency drift [30–32]	Non-detection zone issues and balanced islanding circumstances can be resolved.	Microcontroller-based implementation is easy.
Impedance measurement [33,34]	The functioning of impedance measurement-based ID is beneficial due to the lack of NDZ.	The operation is not possible in the case of a parallel inverter operation
Harmonic signal injection [35,36]	It is advantageous for cases where a power balance is needed between generation and consumer operation.	Some identification delay may be present in the algorithm.
Slip mode frequency shift [37,38]	Small NDZ makes the algorithm highly effective.	Introduce a phase shift parabola, which may result in signal noise and measurement error.
Sandia frequency shift [22,39]	Easy implementation due to the presence of small NDZ.	Power quality and system stability are still concerns that might result in unfavorable system behavior.
Sandia voltage shift [40,41]	When compared to other positive feedback-based ID techniques, it has an extremely rapid detection speed.	The strategies may have a limited impact on the system's power quality and transient responsiveness.
Frequency jump [42,43]	Effective for several DGs that do not operate in tandem.	
Virtual capacitor and inductor [44]	Parallel operation reduces efficiency.	Simple and effective ID technique with decreased output harmonics.

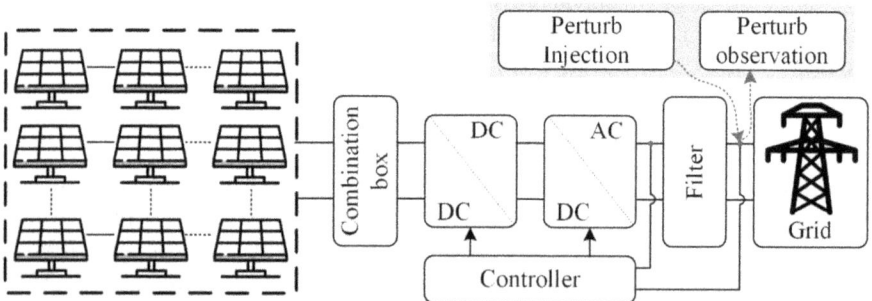

FIGURE 7.2   A schematic diagram for the active ID approach using a PV system.

sequence current is fed into the voltage source inverter of DGs in reference [26]. The approach can identify many DGs working simultaneously; however, it has a flaw with NDZs that delays detection by 60 ms. The use of an active IDM to conduct quick ID by injecting an adjustable pulse current into the grid is then detailed in reference [24]; however, the detection time and rise in voltage level are two of the method's key drawbacks.

7.2.2.2 Active Frequency Drift

The continuous drift in the fundamental frequency causes distorted output frequency in the active frequency drift IDM. The system tends to match the biased frequency to the fundamental frequency by adjusting the frequency after each half cycle followed by dead time. The current in each cycle is presented as:

$$i = \sqrt{2I} \, \sin\left[2\pi\left(f_v + \delta f\right)\right] \tag{7.4}$$

where, i is the output current, I denotes the maximum current, f_v denotes the voltage frequency, and δ_f denotes the frequency change.

The active frequency shift is investigated in reference [30] by quantifying chopper fraction values based on the THD permissible limit. Positive feedback is also added to minimize the NDZ and increase efficiency. However, it is quite doubtful that this technology will be implemented for multi-inverter operation. In reference [31], a voltage pulse perturbation is used to accomplish active frequency detection. Regardless of the number of DGs linked in the grid, this associated approach identifies islanding without adjusting the threshold.

7.2.2.3 Impedance Measurement

The voltage at PCC and its effect on the current are measured using this approach. Changes in voltage have an influence on current in a grid-connected DG is given as

$$\Delta V = \frac{\Delta P_{\mathrm{DG}}}{2} \sqrt{\frac{R}{P_{\mathrm{DG}}}} \tag{7.5}$$

where P_{DG} denotes the active power of DG, R represents the resistive load, and V represents the voltage at PCC. The IDM has a small NDZ, and even though the inverter power output and load are regulated, the inverter output varies, resulting in false activation of trip signal [46]. In reference [34], the grid-connected inverter operates as a virtual impedance, with a frequency that varies slightly from the grid's fundamental frequency. As a result, the frequency and amplitude of the local load will diverge from the nominal value in the event of an abnormality. When using several inverters, this approach is inefficient since these fluctuations might lead to erroneous ID and disconnection.

7.2.2.4 Harmonic Signal Injection

The inverter of DGs injects ninth-order harmonics with the current into the grids in the harmonic signal injection-based IDM. The distortion factor must be less than 4%.

In the case of pre-existing 9th order harmonics in the system, particularly for multi-inverter operation, there is a risk of faulty ID and disconnecting of the DGs from the grid [47]. To address the aforementioned concerns, algorithms that can select any of the harmonics and even analyze two separate harmonics have been developed as discussed in reference [48]. Additionally, in reference [36], To create harmonic current disruption, the authors advise that the inverter's parallel equivalent impedance be altered.

7.2.2.5　Slip Mode Frequency Shift

When the system is aberrant, positive feedback tends to destabilize in this IDM. The solution may be implemented by making a little change to the current phase lock loop (PLL) filter, which automatically phases out the system whenever an irregularity is detected [18]. In reference [38], The response of slip modes ID for various inverter control methods was investigated, and it was discovered that the effectiveness of the IDM is considerably greater with the current controlled strategy than with a constant power control scheme.

7.2.2.6　Sandia Frequency Shift

The Sandia frequency sit method attempts to overcome the flaws in the slip-mode-based frequency modulation technique. In positive feedback, the error acquired from the dead zone is referred to as a frequency error. Because the DGs are connected to the grid, a change in frequency might cause a trip; however, the grid will attempt to maintain the system stable. The feedback equation is as follows:

$$\theta = \frac{\pi}{2}\left(cf_o + K\left(f - f_o\right)\right) \tag{7.6}$$

where the terminal and fundamental frequencies are denoted by f, and f_o, respectively. K denotes the positive-feedback gain, and chopping frequency is signified as cf_o. During the abnormal condition of operation, the frequency error grows along the dead zone, presenting a benefit over the typical frequency shift approach. The efficiency of transient response is significantly lowered when a high-density source is used.

In reference [22], when compared to traditional Sandia frequency shift methods, an artificially immunized Sandia frequency shift methodology is presented to obtain lower THD. Furthermore, because the suggested technique requires the use of a current control-based inverter, it is determined that this method is not suited for small DGs [49].

7.2.2.7　Sandia Voltage Shift

During the ID mechanism, the Sandia voltage shift is a positive-feedback IDM in which the inverter power declines along with the voltage. When the threshold value is surpassed, the feedback control detects the loss in power and trips the power relay. In reference [41], by upgrading the positive feedback with exponential product modification, a modified Sandia voltage shift is suggested that decreases the NDZ and achieves a quicker ID. Due to the disturbance put into the system, the power quality could still have an influence.

7.2.2.8 Frequency Jump

Unlike the frequency biased approach, in which the dead zone is missing at periods, the frequency jump IDM adds a dead zone after each cycle. The inverter current at the DG-grid connector is changed, and the grid connection voltage at PCC is measured. During the islanding state, the current and voltage values are adjusted according to the inverter operation [30]. The frequency modification assists in the detection of the operation's islanded status. Numerous interconnections at the PCC have an impact on this approach, lowering the IDM's efficiency.

7.2.3 Remote Islanding Detection Technique

Since the DGs interact directly with the grid, it's among the most effective and convenient modes of ID. The difficulty of execution and expense of setup are the method's key drawbacks. Table 7.3 describes a couple of the most used IDMs, and Figure 7.3 depicts the arrangement.

7.2.3.1 Programmable Logic Controller

The programmable logic controller (PLC) is a communication signal-based controller that is used in systems with a power rating of 25 kV or greater. The power system must run at the fundamental frequency or below due to the low-pass nature of the filter [57]. DGs receive a signal assuring connection between DGs and the grid during regular operation. A cut-off signal is issued to the circuit breaker to separate the DGs from the grid during the abnormal state. The approach is extremely reliable and simple to use.

In reference [52], a quick overview of PLC-based ID has been provided. Furthermore, in reference [51], To identify any irregular grid functioning, a sub-harmonic signal is inserted into a power line. When compared to the voltage injection approach, this method improves system flexibility and allows for the use of a compact transformer. The disadvantage of this technique is that in the event of

TABLE 7.3
Overview of Different Remote ID Techniques

Method	Merit	Demerit
Programmable logic controller [50–52]	It interacts in real time with all DGs and is extremely dependable and accurate.	Implementation and maintenance costs are too expensive.
Phasor measuring units [53,54]	The detecting technique is simple to deploy because no extra gadget is required. For time synchronization stamping of phasor measurement in DG systems that are geographically distributed, a global positioning system (GPS) may be required.	Not strong enough to handle a variety of signal kinds across the network.
Transfer trip [55,56]	With a modest NDZ, it is a straightforward idea to implement.	It is both costly and difficult to deploy and maintain.

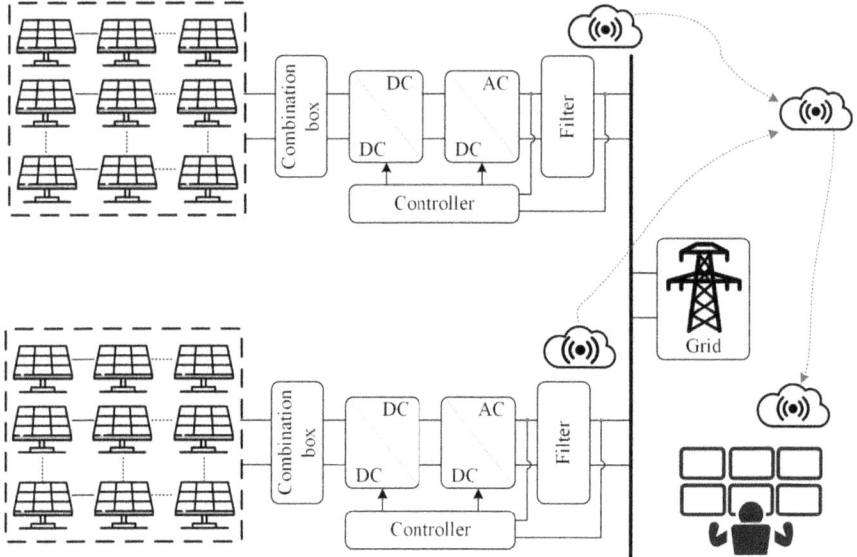

FIGURE 7.3 Brief layout for the remote ID technique with PV systems.

non-radial signals, numerous signals must be conveyed, which raises the implementation cost [58].

7.2.3.2 Phasor Measuring Units

The synchro-phasor measurement is employed in this IDM to identify the DGs' working conditions. The measurement sensors were specific to each substation, and the bulk of the monitoring and analysis is based on the measurement's accessibility. Furthermore, the phase measuring unit (PMU) contributes to achieving quicker data transfer point-to-point connectivity [59]. Because the DGs are used in limited geographical regions, the shift in the angle caused by the voltage phases may be reliably detected using PMUs [60].

In reference [53], To improve the PMU's capabilities, a comprehensive ensemble for data analytics is undertaken. At the same time, the technique reduces computing effort and improves grid observability for PMU. Further in reference [54], before being communicated over PMUs, the data is analyzed using a machine learning classifier. This method boosted ID speed, but it had limitations in terms of power system flexibility and PMU losses.

7.2.3.3 Transfer Trip

The transfer trip method depends on a centrally controlled DG controller that keeps track of the status of all circuit breakers in the system. The networks are based on grid anomaly detection and coordinated operations. The circuit breaker tripping sends signals to the DG controller, which account for the islanded region and calculates the future operation depending on the available load [61]. The transfer trip does have a benefit over PLC in that each DG may be monitored separately throughout the radial connection [62].

In reference [55], The resonant condition, as well as active power matching, are used to suggest a low-cost transfer trip approach. The current in the circuit breaker, as well as the frequency and voltage values, are monitored in accordance with the threshold. Furthermore, a direct transfer trip is proposed in reference [56] to overcome the SCADA system's limitations. The implementation of a transfer trip for a major power system, on the other hand, adds to the complexity and requires a significant expenditure.

7.2.4 HYBRID ISLANDING DETECTION TECHNIQUE

A hybrid IDM is a system that combines two or more current ID algorithms to surpass each other's limitations and deliver accurate and fast identification. Both passive and active IDMs may be found in it. Table 7.4 details a few of the most extensively used hybrid algorithms in the literature, and Figure 7.4 illustrates the arrangement.

TABLE 7.4
Overview of Different Hybrid ID Techniques

Method	Merit	Demerit
Voltage and reactive power shift [63–65]	The method increases the system's reliability and fault tolerance.	The issue of power quality and system stability persists.
SFS and Q-f based scheme [66,67]	It offers quick identification as well as load power factor improvement and voltage regulation.	The system's power quality, as well as a threshold detection issue, might be a problem.
Positive feedback and voltage unbalance [68,69]	Reduce the number of false positives and trips.	It is not possible to implement in a single-phase system.
SFS and ROCOF [70]	The approach may be used in a multi-DG system.	In some cases, determining the trip border might be difficult.

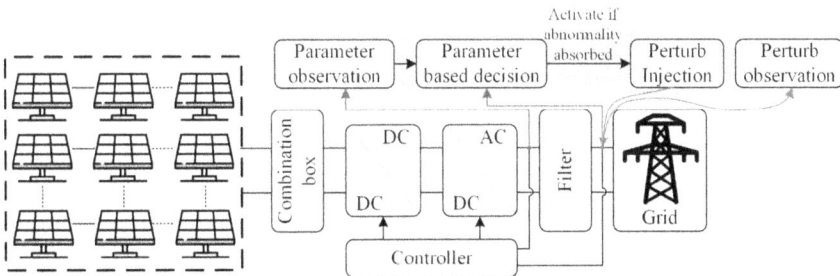

FIGURE 7.4   Brief layout for the remote hybrid ID technique with PV systems.

7.2.4.1 Voltage and Reactive Power Shift

This approach analyzes the reactive power of a grid-connected system to detect changes in reactive power, which helps determine the system's voltage fluctuation [65]. Using a synchronous rotating frame of reference, the searching method is implemented inside the inverter's voltage control unit [64].

7.2.4.2 SFS and Q-f-Based Scheme

The variation in voltage, as well as the variation in reactive power and frequency, is measured in this method. This IDM assists in the monitoring of the grid-connected mode's dynamic functioning and decreases the time it takes to notice abnormalities [67]. Furthermore, one of the strengths of its implementation is the technique's multi-DG applicability [66].

7.2.4.3 Positive Feedback and Voltage Unbalance

The major goal of this strategy is to prevent false ID and disconnection among grid-connected DGs. The voltage imbalance factor, which enhances detection accuracy, is one of the method's main needs [68,69]. Additionally, during PCC, positive feedback is used to monitor current variations.

7.2.4.4 SFS and ROCOF

To achieve efficient ID in the SFS and ROCOF methods, a mix of active and passive IDMs is devised [71]. The passive IDM will first test the system for anomalies. During abnormal operating conditions, the active IDM is engaged to cross-check the situation after the passive IDM has discovered the issue, before generating the trigger signal to disconnect the DGs from the utility.

7.2.5 DATA-DRIVEN ISLANDING DETECTION TECHNIQUE

All the IDMs covered in the preceding sections rely on real-time data or the sensor's pre-defined limitations. They are usually correct, but when there are some abnormalities, the pre-set data will not be enough, resulting in an accidental trip. As a result, data-driven techniques to ID are utilized to overcome this constraint. Using data from multiple phases of operation, this strategy trains the detection algorithm by finding unique patterns in the system's output characteristics. The decision-making method is also carried out based on these taught patterns. Table 7.5 summarizes the most often used data-driven IDMs in the literature, and Figure 7.5 depicts the arrangement.

7.2.5.1 Intelligent Approach

The algorithm is taught to imitate replies that are comparable to human intelligence in this ID method. Depending on the operation at hand, these approaches follow a certain set of predetermined criteria. The change in the input data activates the output, which is based on the ruleset, and actions are taken. In reference [105], Three input variables are given into the system, and the output indicates whether the system is defective or not based on linguistic fuzzy rules. Similarly, in reference [106], for the same mode of operation, 11 inputs are examined. As the number of input variables grows, the intelligent controller's efficiency grows, but it also becomes more

TABLE 7.5

Overview of Different Data-Driven ID Techniques

Method		Merit	Demerit
Intelligent approach [72–79]		Good accuracy and the capacity to manage GC DG with multiple inverters.	The outcome is abstract and based on a collection of specified rule sets.
Learning-based approach [80–85]		It's simple to classify the many phases of operation, and it works with several DG systems.	It is challenging to build and compute since it requires a huge database for training.
Data pre-processing approach	Wavelet transform [86–90]	The ID may function at multiple resolution bands due to the changeable size of the time-frequency window.	It is extremely sensitive to noise signals and operates mostly in the low-frequency range.
	S-transform [91–94]	Its local spectral features are remarkable in that they integrate referenced local phase information with frequency-dependent time-frequency space resolution.	Its shortcomings include a long computation time and smaller gaussian windows.
	Mathematical morphology [95,96]	Adapts time-domain analysis to efficiently denoise the data obtained.	Its action on a single structuring element only supports single-direction characteristics and is incompatible with data that are randomly orientated.
	Hilbert-Haung transform [97–99]	For both nonstationary and nonlinear data analysis, creates potentially feasible energy representations in time-frequency space.	The empirical model distribution of the transform approach is incapable of quantitatively dissolving components with frequency proportions close to unity.
	Principle component analysis [100,101]	Removes associated features, minimizes data overfitting, and enhances data presentation in time-frequency space.	Independent variables are becoming less open to interpretation, resulting in data loss.
	Gauss-Newton algorithm [102,103]	With relative efficiency, it provides a greater radius for data convergence and tighter error estimates on distances for each feature data.	With inadequate starting estimations, this strategy may result in the detection system with the lowest training and testing accuracies.
	Phaselet algorithm [104]	The variable data window for computing the phasor allows for a quick estimate while utilizing a limited sample set.	Because of the changing window size, it has a problem with undesired categorization during transients in the data.

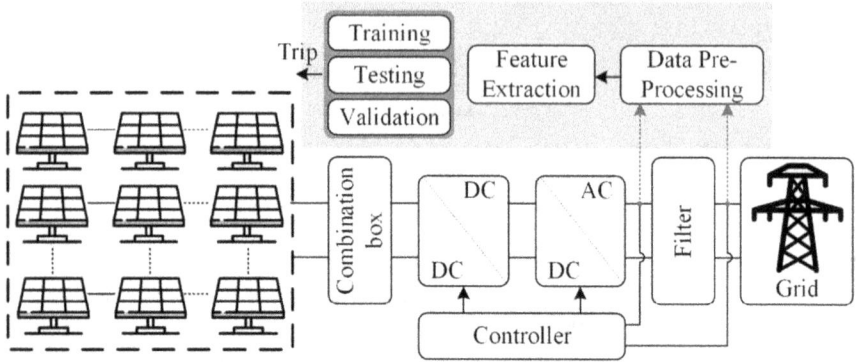

FIGURE 7.5   Brief layout for the remote data-driven ID technique with PV systems.

complicated, requiring more computer resources. Furthermore, for accurate classification, the construction of an adaptive neuro-fuzzy inference system (ANFIS) [107] for nonlinear inputs require fewer inputs, lowering both time and computing complexity. In comparison to a fuzzy-based controller, it can also handle unclear data considerably more effectively. As a result, in the case of a multi-layer DG system, it is usually employed for islanding condition detection [79].

7.2.5.2 Learning-Based Approach

The historical database is utilized to train a classifier in this method, which results in the accurate detection of operational states. As in most cases, the method is implemented in a feed-forward network, with the electrical signal from the PCC serving as an input [108]. To improve the accuracy of identification, the signal is pre-processed to features extracted [109]. Once the features have been retrieved, the learning algorithm distributes a certain quantity of data at random for testing, training, and validation. When the learning process is complete, the algorithm is given a random value to identify the functioning state [83,110,111].

7.2.5.3 Data Pre-Processing Approach

Before the operational condition is categorized, the electrical signal collected from the sensor at PCC is even further processed in this method. Different signal processing methods, which are detailed below, are used to pre-process the signal.

- **Wavelet Transform:**
     The magnitude of the signal coefficients highlights signal abnormalities, and the wavelet transformations identify the system's islanding situation [86]. In general, discrete wavelet transformations are used to shift and dilate the signals in a time-frequency spectrum utilizing high-pass and low-pass filters. This method [89] discretizes the signal into various sample frames as in time-frequency bandwidth and detects anomalies that may be compared to the system's normal operation. The constraints of an NDZ with traditional ID techniques are solved by this process.

- **S-Transform:**
  The S-transform is a customized discrete wavelet transform with two zero moments and altered temporal localization. The S-transform, unlike the wavelet transform, is based on the construction of many filters, giving it a greater degree of freedom in scaling the signal [94].

  This gives the S-transform an edge in determining the sensitivity indices' distinctive qualities throughout the feature extraction phase [91]. The retrieved characteristics can also be utilized to identify between normal, fault, and other types of disturbances.

- **Mathematical Morphology:**
  Based on the signal structure and set theory, this method determines the spatial patterns of the time-domain signal. It also employs a number of morphological operators in which a structural element is employed as an investigation to extract a signal's significant aspects [95]. The form of these structural operators is usually determined by the structure of the signal. The dilation and erosion procedures for inflating and compressing data, as well as the opening and closure operators for smoothing and filtering the data's sharp structures, are the core operations of mathematical morphology. Furthermore, when paired with other operators, this technique can effectively identify islanding and non-islanding circumstances in the network.

- **Hilbert-Haung Transformation:**
  An empirical model decomposition technique is used to generate the Hilbert-Huang transformation. This method decomposes signal data into single component signals known as intrinsic mode functions, which must meet a series of constraints throughout the decomposition process. These criteria specify that the signal should have an equal number of extrema and zero crossings for each packet of local maxima and minima at each location, as well as a zero average value for each envelope of local maxima and minima [99]. On decomposition, the Hilbert transform is used to retrieve the signal's characteristics from each intrinsic mode function [97]. Under a high-power imbalance, these feature vectors may be further examined for multiple islanding and non-islanding scenarios.

- **Principle Component Analysis:**
  Under a high imbalance of power, these feature vectors may be further examined for multiple islanding and non-islanding scenarios. The principal components are calculated in the order of decreasing variability, with the first principal component containing the original signal's substantial fluctuation [101]. Furthermore, by doing eigenvalue/eigenvector analysis, the principal component analysis orthogonally spins the original data to project the largest variations onto a new axis [100]. The islanding/non-islanding circumstances with less NDZ and a better level of confidence are further identified by these maximal variations.

- **Gauss-Newton Algorithm:**

     The Gaussian-Newton approach estimates parameters by minimizing the difference between estimates by solving nonlinear least-square problems. By preserving the high dynamic precision of the measurement, this approach iteratively decreases the error and recognizes the signal fluctuations within a few sample intervals [103]. Furthermore, when paired with iterative and zero-crossing approaches, the Gaussian-Newton method increases the iteration convergence rate [102]. This aids in obtaining high precision and monitoring speed for modifications in the power system network's islanding/non-islanding operation.

- **Phaselet Algorithm:**

     A partial accumulation of the product of sample data over a sample window is defined as phaselets. The mathematical meaning of these sample windows is unrestricted, and they can successfully estimate the phasors across the window length. In addition, the phaselet technique uses a random window-length filtering approach to create phasor estimating progress for ID. The phaselets function on a fixed period and window while the power system is in normal operating conditions, but when the system is in an islanding situation, the phaselet's filter window size is automatically reduced [103]. The islanding/non-islanding events in the system are clearly identified by the fluctuation of window size in relation to the fixed half or full cycles.

7.3   FAULT RIDE-THROUGH

Before the trip signal is created, the system attempts to recover from of the problem after it has been discovered. The recovery process has been described in detail in the preceding sections. The national grid code [112] and IEC 62112 [113] are used to determine the range of operations at which the system will run smoothly, and the control will strive to restore the system in the event of a fault. If the problem is not fixed within a certain amount of time, a trip signal is generated, which disconnects the PV from the grid. A basic overview of the grid code is shown in Figure 7.6.

     Low-voltage ride-through (LVRT) and dynamic voltage support (DVS) capabilities are two control modes that explain state-of-the-art needs and settings according to international standards and grid regulations. The general purpose is to think about control modes, specifically voltage control modes that affect short-term voltage dynamics. For the reaction of the PV system to grid problems, especially voltage sag, three alternative control mechanisms are examined.

7.3.1   No Low Voltage Ride-Through Capability (No LVRT)

This mode makes advantage of the PV plant's under-voltage protection mechanism. When the voltage falls below 6% of rated voltage for more than 2 seconds, the PV inverter decouples from the grid and disconnects from the grid. The root-mean-square values are derived from the momentary readings and transmitted to the under-voltage protection function, which sets the voltage restrictions and time delays. The OR function receives the tripping signal and opens the circuit breaker, disconnecting the PV plant.

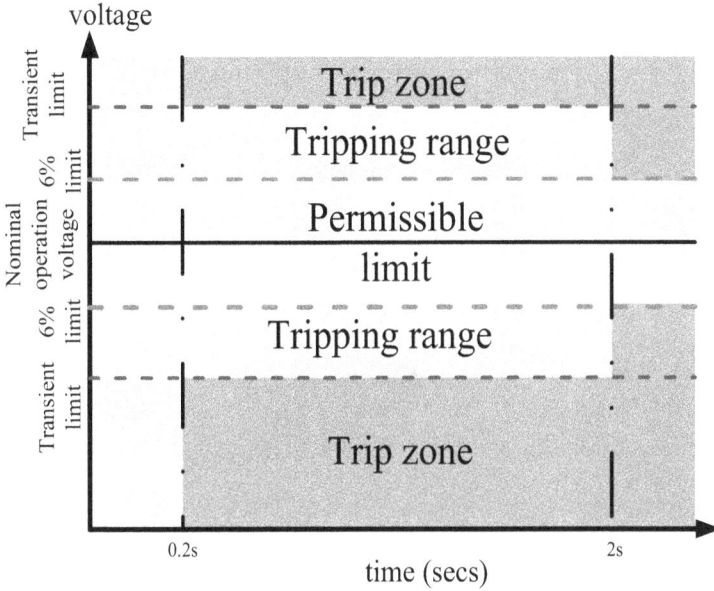

FIGURE 7.6   Tripping range as per IEC 62116.

7.3.2 Low Voltage Ride-Through and Blocking Mode

This control mode takes into account two factors: the LVRT and the PV system's blocking capabilities. The LVRT feature, or FRT capability in general, allows the PV system to stay grid – connected during power drops. Figure 7.7 depicts the LVRT feature under consideration.

If the voltage falls inside the no-trip zone during the fault-on time, the PV system must remain grid – connected and ride through the fault. The PV system, on the other hand, might be disconnected if the voltage falls underneath the LVRT characteristic. An incremental under-voltage protection system containing n functions can be used in reality to provide the LVRT capabilities. The reactive and active current behavior during the fault-on time is the second component of this control style. During the fault-on time, blocking implies that the PV system doesn't always insert any active or reactive current. Many inverters use the blocking mode, which is also known as zero power or brief stoppage mode [114].

7.3.3 Dynamic Voltage Support

The DVS feature, also known as dynamic reactive power support or dynamic voltage control, is enabled by this component of the control mode. DVS defines a reactive current injection that is proportional to the voltage variation. In the following sections, we will go through the specifics of establishing DVS using active and reactive power regulation.

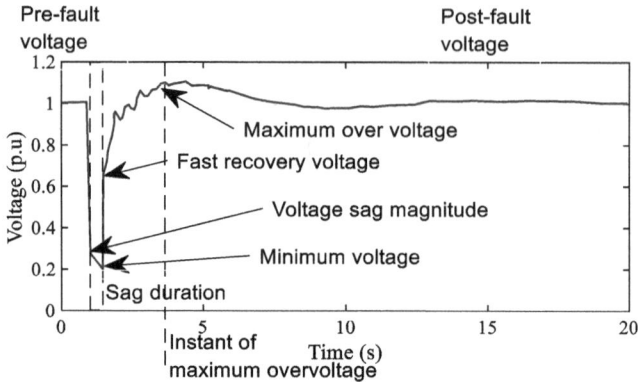

FIGURE 7.7   Characteristics of fault-induced voltage sag.

7.3.4 ADDITIONAL REACTIVE AND ACTIVE CURRENT CONTROL

The goal of this research is to create extra-reactive and active current control (ARACC) to provide an efficient low voltage ride-through with no transients in system operation and within the grid standards' time constraints. In general, the X/R ratios in LV networks are higher than those at higher voltage levels. This indicates that active power in the LV network has a greater influence on voltage than it does otherwise. The angle at which extra current is injected may be changed. The angle is chosen at the impedance angle at the common coupling point in this study (PCC). Because the X/R ratio at the PCC is smaller, this results in greater efficacy and stability than pure reactive current injection. The key benefit of ARACC was a greater active power in-feed from DG right after the fault was cleared, which might be useful in low-inertia power systems.

- Reactive current control

    Initially, the traditional reactive current control is addressed to build the ARACC control technique. Figure 7.8 depicts the controller design process for reactive current control.

    The following are the equations that correlate to RCC:

$$I_{\text{Flt}} = \bar{I}_0 + \Delta I(\Delta V) = I_{d,\,\text{Flt}} + jI_{q,\,\text{Flt}} \qquad (7.7)$$

where I_{Flt} is the current at the PCC during a fault, $\bar{I}_0$ is the pre-fault current phasor magnitude, ΔI is the additional current injection during fault mode, ΔV is the voltage deviation from the pre-fault value, and $I_{d,\,\text{Flt}}$ and $I_{q,\,\text{Flt}}$ are direct and quadrature axis current phasor magnitude during fault mode, respectively.

Furthermore,

$$\Delta I(\Delta V) = e^{j\left(\phi_G + \frac{\Pi}{2}\right)} \cdot \begin{cases} 0, |\Delta V| \leq V_{\text{DB}} \\ k_{\text{RCC}} \cdot [\Delta V \mp V_{\text{DB}}], |\Delta V| > V_{\text{DB}} \end{cases} \qquad (7.8)$$

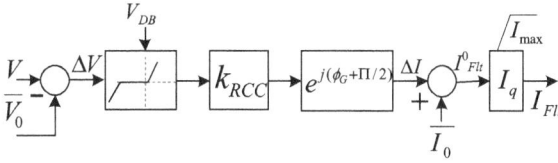

FIGURE 7.8 DVS through reactive current control mode.

where ϕ_G is the voltage angle of grid equivalent, k_{RCC} is the controller gain for RCC, and V_{DB} is the voltage deadband for fast voltage control.

$$\Delta V = V - \overline{V_0} \qquad (7.9)$$

where V is the voltage and $\overline{V_0}$ is the average voltage phasor value.

$$\overline{I_0} = \frac{1}{T} \int_{t'-T}^{t} I(t')dt' \qquad (7.10)$$

where T is the time constant, t is the time in [s], t' is the time in p.u. and I is a current phasor.

$$\overline{V_0} = \frac{1}{T} \int_{t'-T}^{t} V(t')dt' \qquad (7.11)$$

and with current limitation

$$I_{q,\,Flt} = \overline{I}_{q,\,0} + \Delta I_q = \min^{I_{max}}\left(\overline{I}_{q,\,0} + k_{RCC}\cdot\left[(\Delta V)\mp V_{DB}\right]\right) \qquad (7.12)$$

where I_q is the quadrature axis current phasor magnitude during the pre-fault mode, ΔI_q is the change in the value of quadrature axis current component, and I_{max} is the maximum current.

$$I_{d,Flt} = \begin{cases} 0 & \text{, for } i_{FRT\_CI\_PRIO\_MOD} = 0 \\ \left(I_{max} - I_{q,Flt}\right) & \text{, for } i_{FRT\_CI\_PRIO\_MOD} = 1 \\ \min^{I_{max}^0}\left(\sqrt{\left(I_{max}\right)^2 - \left(I_{q,Flt}\right)^2}\right) & \text{, for } i_{FRT\_CI\_PRIO\_MOD} = 2 \end{cases} \qquad (7.13)$$

where $i_{FRT\_CI\_PRIO\_MOD}$ is the current priority mode: $1 = $ arithmetic (abs), $2 = $ geometric (sqrt), other $=$ set other value to zero.

$$I_{max}^0 = \begin{cases} I_{max}, \text{ for } i_{FRT\_CI\_STAB} = 0 \\ \overline{I}_{d,0} \cdot (v_{pv})^2, \text{ for } i_{FRT\_CI\_STAB} = 1 \end{cases} \qquad (7.14)$$

where v_{pv} is the voltage output of the PV system and $i_{FRT\_CI\_STAB}$ is stability improvement during fault by voltage-dependent i_d reduction: $0 =$ no; $1 =$ yes.

- Reactive and active current control

    By altering the impedance angle at the PCC, the PV inverter's last fault control method is to inject extra-reactive and active current (ARAC). Figure 7.9 shows the block diagram for the suggested manner of operation.

    From equation 7.14, the mathematics corresponding to reactive and active current control are given as

$$I_{Flt}^0 = \overline{I}_0 + \Delta I(\Delta V) = I_{d,Flt}^0 + jI_{q,Flt}^0 \qquad (7.15)$$

$$I_{Flt} = I_{Flt}^0 \cdot F_{max} = I_{d,Flt} + jI_{q,Flt} \qquad (7.16)$$

where 0 in the superscript defines a non-limited quantity and F_{max} is a scaling factor for current limitation during fault control mode with

$$\Delta I(\Delta V) = e^{j(\phi_G + \Psi_{ARACC})} \cdot \begin{cases} 0, |\Delta V| \le V_{DB} \\ k_{RACC} \cdot [\Delta V \mp V_{DB}], |\Delta V| > V_{DB} \end{cases} \qquad (7.17)$$

where k_{RACC} is controller gain for active and reactive current control.

Here, stability improvement during fault by voltage-dependent i_d reduction: $0 =$ no; $1 =$ yes.

with current limitation

$$I_{q,Flt} = I_{q,Flt}^0 \cdot F_{max} \qquad (7.18)$$

$$I_{d,Flt} = I_{d,Flt}^0 \cdot F_{max} \qquad (7.19)$$

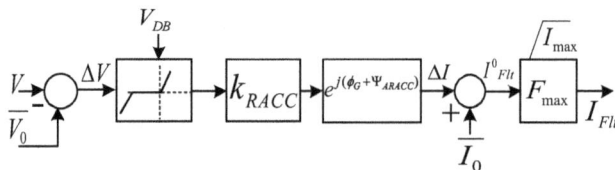

FIGURE 7.9 DVS through additional reactive and active current control modes.

$$I_{q,\,Flt}^{0} = \overline{I}_{q,\,0} + \Delta I_q = \overline{I}_{q,\,0} + \left[k_{RACC}.[\Delta V \mp V_{DB}].\sin(\Psi_{ARACC}) \right] \qquad (7.20)$$

$$I_{d,\,Flt}^{0} = \overline{I}_{d,\,0} + \Delta I_d = \overline{I}_{d,\,0} + \left[k_{RACC}.[\Delta V \mp V_{DB}].\cos(\Psi_{ARACC}) \right] \qquad (7.21)$$

where

$$\overline{I}_{d,\,0}^{0} = \begin{cases} \overline{I}_{d,\,0}, \text{ for } i_{FRT\_CL\_STAB} = 0 \\ \overline{I}_{d,\,0}.(V_{DG})^2, \text{ for } i_{FRT\_CL\_STAB} = 1 \end{cases} \qquad (7.22)$$

$$F_{max} = \max_{1} \left(\frac{I_{max}}{|I_{Flt}^{0}|} \right) \qquad (7.23)$$

The injected increased active and reactive current is represented graphically in Figure 7.10. It can be seen in the diagram that when an active current is inserted during a voltage decrease, the active power often does not rise. In addition, the active sign convection denotes:

- Injection of positive active power $S = V \cdot I^* = P + jQ = (V \cdot I_d - jV \cdot I_q)$
- Pre-fault current: exchange of negative reactive power (positive reactive current, $I_q > 0$) causes voltage decreases (inductive power factor).

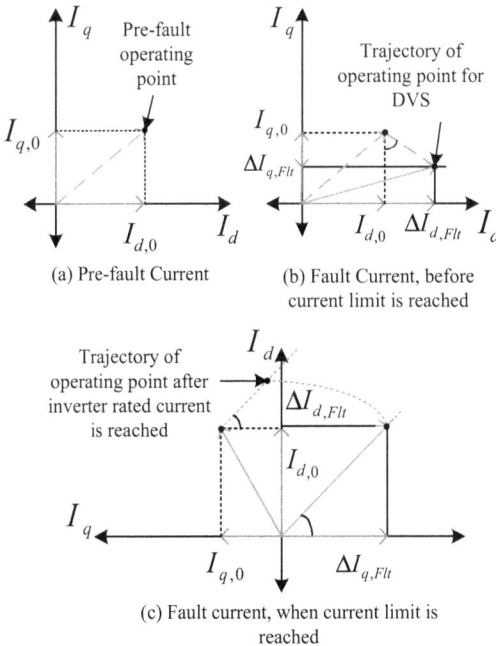

(a) Pre-fault Current

(b) Fault Current, before current limit is reached

(c) Fault current, when current limit is reached

FIGURE 7.10 Scaling of direct and quadrature axis current in additional reactive and active current control mode.

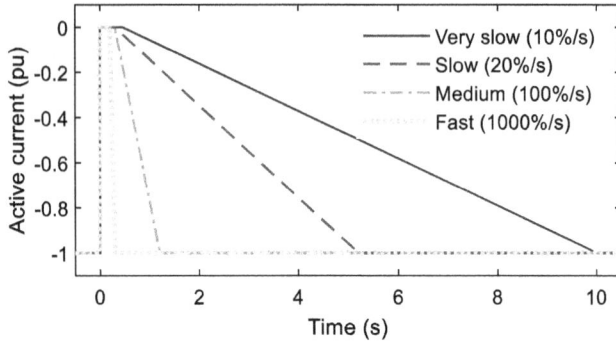

FIGURE 7.11   Active current recovery rate during the post-fault period.

- Full dynamic network support: exchange of positive reactive power (negative reactive current, $I_q < 0$) causes voltage increase (capacitive power factor).

• Active Current Recovery Rate

The active current recovery rate during the post-fault phase is used to calculate how long it will take for the PV system's active power to be recovered following a fault. Figure 7.11 depicts the recovery rates under consideration, which are categorized as extremely slow, moderate, medium, and quick. During the extremely slow and recovery modes, the active current of a PV plant is restored at a rate of 10%/sec and 20%/sec, respectively [114]. However, the recovery rate in the medium and fast recovery modes is 100%/ sec and 1000%/ sec, respectively. These are the lower limitations set for failures lasting more than 140 ms [115].

7.4   TRANSITION CONTROL

Growing environmental concerns about traditional power plants, combined with rising power consumption, have prompted a shift to DGs system based on renewable sources. Photovoltaic (PV) systems are one of the most widely adopted renewable energy sources, having a large installed base and rapid growth rates. However, their reliance on daily and seasonal oscillations resulted in substantial output fluctuations. Because of this, PV grid integration is a difficult task. To address this, solutions to enhance the electrical network have been proposed as discussed in reference [116]. However, this comes at a significant cost, and when energy is carried across large distances, its overall efficiency suffers [117]. Furthermore, the grid codes for several related countries must be harmonized, which is a lengthy process [118]. Installing distributed energy storage units (ESU) [119] is another option. The high-level control can establish groups that can operate entirely or together as the main power grid and behave similarly to a traditional power plant using communication between the generating, loads, and storage systems [120,121]. Using reduction criteria for the specified priority, the generation-consumption can be optimized. Stationary energy storage is a complementary approach that can be tailored to the specific needs of various services. The demands of the load profile can be addressed with the aid of the (ESU) [122] as well as the energy management system (EMS) [123] while taking into consideration the mission profile. The DGs can also be used in standalone (SA) as well as grid-connected (GC) modes [124]. When the DG is islanded, it is detached from the power system and switches to the SA mode. Local loads are met by DGs and ESUs in this case, based upon their mission profile as well as state of charge (SOC) [125], respectively. Furthermore, based on the load profile, accessibility of PV power, grid, and ESU SOC, the EMS guarantees interaction between DGs, storage devices, and utilities [126]. Furthermore, the EMS assists in the system's stable, adaptable, and secured operation [127]. However, the EMS's biggest flaw is its inability to respond to load dynamics [128].

The limitations of EMS are addressed in this study by recognizing distinct operating modes and proposing a suitable control technique to meet the load demand. In addition to the load profile, the influence of the mission profile upon DG power generation and ESU SOC are considered when creating the controller. Furthermore, in the event of a grid disturbance, the islanding situation is recognized and categorized using machine learning algorithms employing distinct signal characteristics recorded at PCC. The inverter controller is used to sustain the network by injecting reactive power, depending on the severity of the fault. If the recovery is not accomplished within the timeframe specified by the grid code [113], the grid is detached, and the system enters SA mode. When dealing with ESUs in the system, the transition from GC to SA mode and vice versa must be carefully addressed since battery draining may cause a substantial surge in DC link during the transition from GA to SA mode [129].

In this study, a coordinated control method is developed for a two-stage single-phase grid-connected PV system that can achieve power management across PV, ESU, and grid, as well as a smooth transition between SA and GC modes. Depending on the mission profile, battery SOC, and load profile, the power management approach coordinates the PV boost converter, ESU bidirectional DC–DC converter, and full-bridge inverter functioning with the grid. In addition, for the transitions between GC and SA modes, a closed-loop control method is designed that considers the capacitor current feedback and the outside current loop.

7.4.1 Control Structure for Inverter Mode Transition

Massive reactive current circulation between both the PV system as well as the grid will occur throughout the changeover period.

A control mechanism is provided to efficiently accomplish the transition depending on the synchronization criteria. The grid voltage (V_g) as well as the voltage on the PV system side directly after the transformer (V_i) are both considered in Figure 7.12. For a smooth transition, the value of V_i must be identical to the value of V_g. It is necessary to guarantee that output current i_2 is zero and that V_i equals the capacitor voltage (V_C) before the inverter and grid are connected. As a result, adjusting the capacitor voltage to match the grid voltage for synchronization makes controlling the inverter voltage simple. The sequence matching requirements are followed.

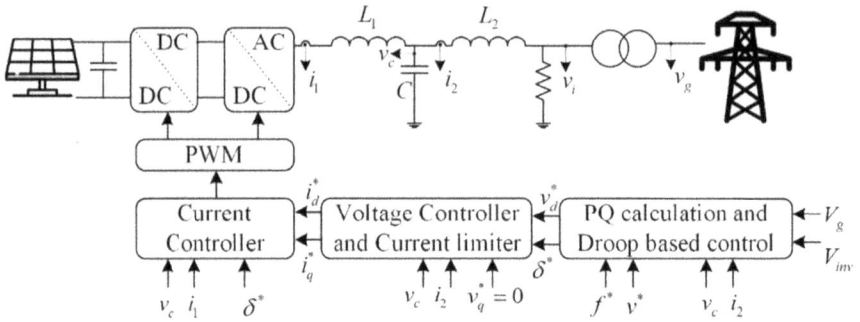

FIGURE 7.12 Modified control structure representation.

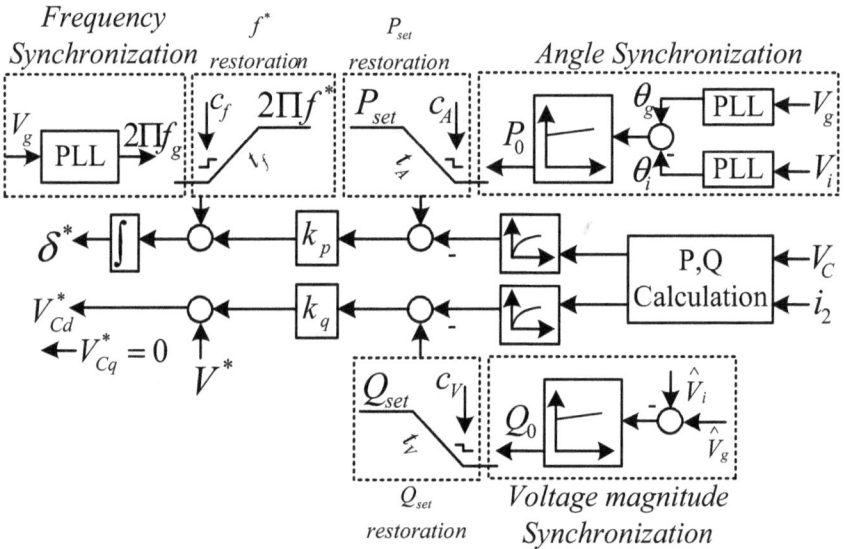

FIGURE 7.13 Control block representation for the proposed controller.

Figure 7.13 depicts the synchronization of voltage, frequency, and angles in greater detail. For frequency matching, a PLL is used, while a proportional integrator (PI)-based controller is used to synchronize voltage amplitude and angle. With ramps of t_f, t_A, and t_V time durations, the restoring frequency and voltage elements (active and reactive) to the target value begins. The duration of the r_{amp} is determined by the grid code. For reactive power and frequency, the appropriate setpoints are attained using the identical firing pulses c_f and c_V. The active power ramping is started with a c_A firing pulse to bring the system to a steady state and distribute the load between the grid and the PV system. The scheme's main goal is to prevent voltage and frequency drops during synchronization and to circulate reactive power to the grid for stability analysis.

7.4.2 Inverter Operating Mode Transition

Certain parameters must be synchronized for a seamless transition from GC to SA mode and vice versa [130].

- PV power generation must have the same frequency as the grid.
- The PV systems and the grid's phase sequences must be the same. If phase sequence matching isn't done correctly, synchronization may fail.
- The PV system's voltage magnitude must match the grids. When the PV system voltage is higher or lower than the grid voltage, it is said to be in over or under excited mode.
- Both the grid and the PV system's phase angles should be in sync. The reference point both for voltages may be seen and supervised.

The transition is carried out in the following manner:

a. *GC to standalone mode*
- If the defective situation persists as well as the grid does not recover within the time limit allowed by the grid code, determine whether the utility is operating in a faulty or normal state.
- At zero crossing of grid currents, the reference current is reduced to zero.
- The gate drive signal at the PCC switch is adjusted from high to low, causing the PV system to be disconnected from the grid.
- The reference voltage (v_{ref}) and the reference grid voltage (v_{gref}) are synchronized.
- During zero crossings, the value of v_{gref} is changed to the value of v_{ref}.
  The controller makes all the modifications of reference current, voltage, and PCC switch from step (b) to (e) at the same time. The PCC switch is activated at positive zero crossings when a problem is detected. As a result, the time it takes to go from GC to SA mode is about equivalent to one utility cycle. According to the evaluation shown in Figure 7.14, the transition between the two modes may be done seamlessly using the flow diagram.

b. *Standalone to a GC mode*
- Identify if the grid is operating in normal conditions.
- The value of v_{ref} is adjusted to match the frequency of the grid voltage. This process is generally required when the grid period is 10 times greater than a reference value.
- A positive zero crossing of output voltage, the value of v_{ref} is shifted to $v_{g\,ref}$ which helps PV system output voltage to attain grid voltage. The time required to complete the process is equal to the utility period.
- When the output voltage of the PV system is equal to the grid voltage then the gate driving signal at the PCC switch is changed from low to high level which presents an interconnection between the PV system and the grid.

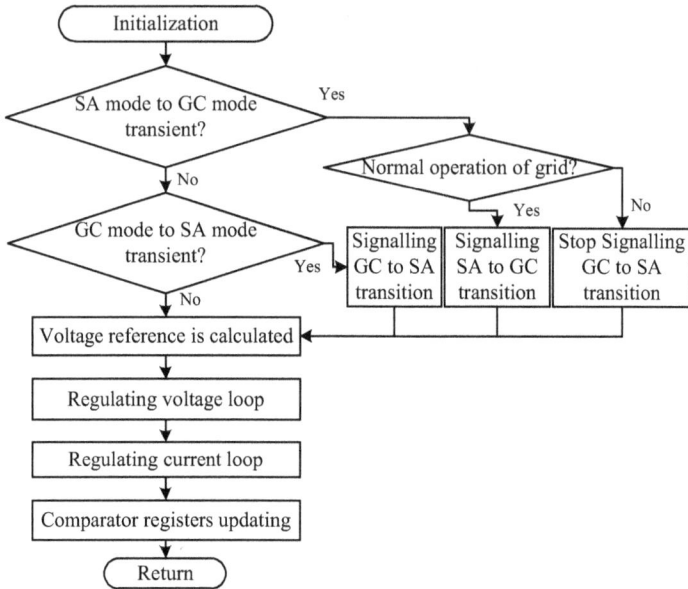

FIGURE 7.14   GC to SA mode of transition.

- The reference grid value is slowly increased to the desired value from zero within a time limit of not more than 6 times that the utility period.

As per the assessment presented in Figure 7.15, the transition can be performed smoothly between the two modes as per the flow diagram.

7.5   SUMMARY

In this chapter, the constant active current reactive power injection strategy for a single-phase GC PV system is designed considering various grid requirements. The developed reactive power injection strategy is motivated to avoid catastrophic failures during grid faults and enhance system reliability and stability during the LVRT process. Furthermore, the design constraints of the reactive power injection strategy were identified and considered by developing the effective constant active current reactive power injection method. The chapter recommends a feasible way of selecting suitable power tools for the new PV inverters by employing the proposed control strategy. The developed reactive power control strategy has been tested with an experimental analysis and the results depicted the effectiveness of the strategy during LVRT operation to sustain the grid voltage. Hence, it can be concluded that Voltage regulation utilizing reactive power compensation with the accurate reactive power control model can be used to fulfil the grid codes in distribution grids with a large PV penetration.

b) Stand-alone to a grid-connected mode

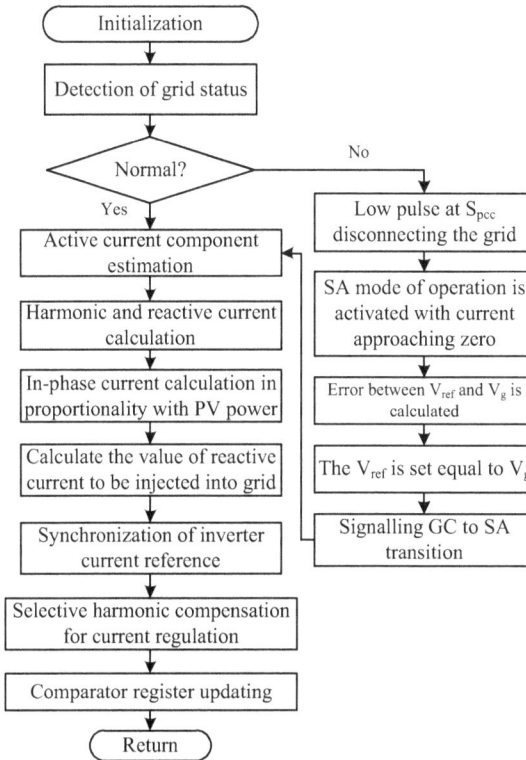

FIGURE 7.15   SA to GC mode of transition.

7.6   QUESTIONS

1. What is islanding?
2. What is passive islanding detection?
3. What is voltage unbalancing in GC solar PV?
4. What is the change in power rates?
5. What is rate of change of frequency?
6. What is over- and under-voltage?
7. What is over and under frequency?
8. How does impedance change when a solar PV system is connected to a grid?
9. What is phase Jump in GC PV systems?
10. What is harmonic distortion in a GC solar PV plant?
11. What is voltage unbalance in GC solar PV plants?
12. What is active islanding detection?
13. What is the importance of active and reactive power in GC PV systems?
14. What are the remote islanding detection techniques?

15. What is a programmable logic controller?
16. What is the Hybrid Islanding Detection Technique?
17. What is the data-driven islanding technique?
18. Which intelligent methods are used for islanding detection?
19. What is a wavelet transform?
20. What is dynamic voltage support?
21. What is the meaning of inverter operating mode transfer?

REFERENCES

[1] J. A. Laghari, H. Mokhlis, M. Karimi, A. H. A. Bakar, and H. Mohamad, "An island-ing detection strategy for distribution network connected with hybrid DG resources," *Renew. Sustain. Energy Rev.*, vol. 45, pp. 662–676, 2015, doi: 10.1016/j.rser.2015.02.037.

[2] M. A. Choudhry and H. Khan, "Power loss reduction in radial distribution system with multiple distributed energy resources through efficient islanding detection," *Energy*, vol. 35, no. 12, pp. 4843–4861, 2010, doi: 10.1016/j.energy.2010.09.003.

[3] M. Bakhshi, R. Noroozian, and G. B. Gharehpetian, "Novel islanding detection method for multiple DGs based on forced helmholtz oscillator," *IEEE Trans. Smart Grid*, vol. 9, no. 6, pp. 6448–6460, 2018, doi: 10.1109/TSG.2017.2712768.

[4] A. G. Abd-Elkader, S. M. Saleh, and M. B. Magdi Eiteba, "A passive islanding detection strategy for multi-distributed generations," *Int. J. Electr. Power Energy Syst.*, vol. 99, no. June 2017, pp. 146–155, 2018, doi: 10.1016/j.ijepes.2018.01.005.

[5] P. Gupta, R. S. Bhatia, and D. K. Jain, "Average absolute frequency deviation value based active islanding detection technique," *IEEE Trans. Smart Grid*, vol. 6, no. 1, pp. 26–35, 2015, doi: 10.1109/TSG.2014.2337751.

[6] Z. Lin et al., "Application of wide area measurement systems to islanding detection of bulk power systems," *IEEE Trans. Power Syst.*, vol. 28, no. 2, pp. 2006–2015, 2013, doi: 10.1109/TPWRS.2013.2250531.

[7] V. R. Reddy and E. S. Sreeraj, "A feedback-based passive islanding detection tech-nique for one-cycle-controlled single-phase inverter used in photovoltaic sys-tems," *IEEE Trans. Ind. Electron.*, vol. 67, no. 8, pp. 6541–6549, 2020, doi: 10.1109/TIE.2019.2938464.

[8] N. Liu, C. Diduch, L. Chang, and J. Su, "A reference impedance-based passive island-ing detection method for inverter-based distributed generation system," *IEEE J. Emerg. Sel. Top. Power Electron.*, vol. 3, no. 4, pp. 1205–1217, 2015, doi: 10.1109/JESTPE.2015.2457671.

[9] X. Xie, C. Huang, and D. Li, "A new passive islanding detection approach considering the dynamic behavior of load in microgrid," *Int. J. Electr. Power Energy Syst.*, vol. 117, no. September 2019, p. 105619, 2020, doi: 10.1016/j.ijepes.2019.105619.

[10] N. Liu, A. Aljankawey, C. Diduch, L. Chang, and J. Su, "Passive islanding detection approach based on tracking the frequency-dependent impedance change," *IEEE Trans. Power Deliv.*, vol. 30, no. 6, pp. 2570–2580, 2015, doi: 10.1109/TPWRD.2015.2418580.

[11] H. Pourbabak and A. Kazemi, "A new technique for islanding detection using volt-age phase angle of inverter-based DGs," *Int. J. Electr. Power Energy Syst.*, vol. 57, pp. 198–205, 2014, doi: 10.1016/j.ijepes.2013.12.008.

[12] S.-I. S. Il Jang and K. H. K.-H. Kim, "An islanding detection method for distributed gen-erations using voltage unbalance and total harmonic distortion of current," *IEEE Trans. Power Deliv.*, vol. 19, no. 2, pp. 745–752, Apr. 2004, doi: 10.1109/TPWRD.2003.822964.

[13] Y. M. Makwana and B. R. Bhalja, "Experimental performance of an islanding detec-tion scheme based on modal components," *IEEE Trans. Smart Grid*, vol. 10, no. 1, pp. 1025–1035, 2019, doi: 10.1109/TSG.2017.2757599.

[14] H. Laaksonen, "Advanced islanding detection functionality for future electricity distribution networks," *IEEE Trans. Power Deliv.*, vol. 28, no. 4, pp. 2056–2064, Oct. 2013, doi: 10.1109/TPWRD.2013.2271317.

[15] S. Raza, H. Mokhlis, H. Arof, J. A. Laghari, and H. Mohamad, "A sensitivity analysis of different power system parameters on islanding detection," *IEEE Trans. Sustain. Energy*, vol. 7, no. 2, pp. 461–470, 2016, doi: 10.1109/TSTE.2015.2499781.

[16] T. Rabuzin, F. Hohn, and L. Nordström, "Computation of sensitivity-based islanding detection parameters for synchronous generators," *Electr. Power Syst. Res.*, vol. 190, no. July 2020, p. 106611, 2021, doi: 10.1016/j.epsr.2020.106611.

[17] W. Xu, S. Martel, and K. Mauch, "*An assessment of distributed generation islanding detection methods and issues for canada*," Canada: N.P, 2004.

[18] M. Ciobotaru, V. G. Agelidis, R. Teodorescu, and F. Blaabjerg, "Accurate and less-disturbing active antiislanding method based on PLL for grid-connected converters," *IEEE Trans. Power Electron.*, vol. 25, no. 6, pp. 1576–1584, Jun. 2010, doi: 10.1109/TPEL.2010.2040088.

[19] C. R. Reddy and K. H. Reddy, "A new passive islanding detection technique for integrated distributed generation system using rate of change of regulator voltage over reactive power at balanced islanding," *J. Electr. Eng. Technol.*, vol. 14, no. 2, pp. 527–534, Mar. 2019, doi: 10.1007/s42835-018-00073-x.

[20] W. Li, Y. Gu, H. Luo, W. Cui, X. He, and C. Xia, "Topology review and derivation methodology of single-phase transformerless photovoltaic inverters for leakage current suppression," *IEEE Trans. Ind. Electron.*, vol. 62, no. 7, pp. 4537–4551, Jul. 2015, doi: 10.1109/TIE.2015.2399278.

[21] Y. Shang, S. Shi, and X. Dong, "Islanding detection based on asymmetric tripping of feeder circuit breaker in ungrounded power distribution system," *J. Mod. Power Syst. Clean Energy*, vol. 3, no. 4, pp. 526–532, Dec. 2015, doi: 10.1007/s40565-015-0162-7.

[22] A. Y. Hatata, E. H. Abd-Raboh, and B. E. Sedhom, "Proposed Sandia frequency shift for anti-islanding detection method based on artificial immune system," *Alexandria Eng. J.*, vol. 57, no. 1, pp. 235–245, 2018, doi: 10.1016/j.aej.2016.12.020.

[23] J. Merino, P. Mendoza-Araya, G. Venkataramanan, and M. Baysal, "Islanding detection in microgrids using harmonic signatures," *IEEE Trans. Power Deliv.*, vol. 30, no. 5, pp. 2102–2109, Oct. 2015, doi: 10.1109/TPWRD.2014.2383412.

[24] C. C. Hou and Y. C. Chen, "Active anti-islanding detection based on pulse current injection for distributed generation systems," *IET Power Electron.*, vol. 6, no. 8, pp. 1658–1667, 2013, doi: 10.1049/iet-pel.2012.0542.

[25] F. J. Lin, Y. S. Huang, K. H. Tan, J. H. Chiu, and Y. R. Chang, "Active islanding detection method using d-axis disturbance signal injection with intelligent control," *IET Gener. Transm. Distrib.*, vol. 7, no. 5, pp. 537–550, 2013, doi: 10.1049/iet-gtd.2012.0488.

[26] B. Bahrani, H. Karimi, and R. Iravani, "Nondetection zone assessment of an active islanding detection method and its experimental evaluation," *IEEE Trans. Power Deliv.*, vol. 26, no. 2, pp. 517–525, 2011, doi: 10.1109/TPWRD.2009.2036016.

[27] G. Hernández-González and R. Iravani, "Current injection for active islanding detection of electronically- interfaced distributed resources," *IEEE Trans. Power Deliv.*, vol. 21, no. 3, pp. 1698–1705, 2006, doi: 10.1109/TPWRD.2006.876980.

[28] S. Murugesan, V. Murali, and S. A. Daniel, "Hybrid analyzing technique for active islanding detection based on d-Axis current injection," *IEEE Syst. J.*, vol. 12, no. 4, pp. 3608–3617, 2018, doi: 10.1109/JSYST.2017.2730364.

[29] H. Karimi, A. Yazdani, and R. Iravani, "Negative-sequence current injection for fast islanding detection of a distributed resource unit," *IEEE Trans. Power Electron.*, vol. 23, no. 1, pp. 298–307, 2008.

[30] M. E. Ropp, M. Begovic, and A. Rohatgi, "Analysis and performance assessment of the active frequency drift method of islanding prevention," *IEEE Trans. Energy Convers.*, vol. 14, no. 3, pp. 810–816, 1999, doi: 10.1109/60.790956.

[31] O. Tsukamoto, T. Okayasu, and K. Yamagishi, "Study on islanding of dispersed photovoltaic power systems connected to a utility power grid," *Sol. Energy*, vol. 70, no. 6, pp. 505–511, 2001, doi: 10.1016/s0038–092x(00)00145–6.

[32] B. Wen, D. Boroyevich, R. Burgos, Z. Shen, and P. Mattaveli, "Impedance-based analysis of active frequency drift islanding detection for grid-tied inverter system," *IEEE Trans. Ind. Appl.*, vol. 52, no. 1, pp. 332–341, 2016, doi: 10.1109/TIA.2015.2480847.

[33] C. L. Trujillo, D. Velasco, E. Figueres, and G. Garcerá, "Analysis of active islanding detection methods for grid-connected microinverters for renewable energy processing," *Appl. Energy*, vol. 87, no. 11, pp. 3591–3605, 2010, doi: 10.1016/j.apenergy.2010.05.014.

[34] W.-J. Chiang, H.-L. Jou, J.-C. Wu, K.-D. Wu, and Y.-T. Feng, "Active islanding detection method for the grid-connected photovoltaic generation system," *Electr. Power Syst. Res.*, vol. 80, no. 4, pp. 372–379, 2010.

[35] K. Jia, Z. Xuan, Y. Lin, H. Wei, and G. Li, "An islanding detection method for grid-connected photovoltaic power system based on Adaboost algorithm," *Diangong Jishu Xuebao/Transactions China Electrotech. Soc.*, vol. 33, no. 5, pp. 1106–1113, 2018, doi: 10.19595/j.cnki.1000–6753.tces.170016.

[36] M. Hamzeh, N. Rashidirad, K. Sheshyekani, and E. Afjei, "A new islanding detection scheme for multiple inverter-based DG systems," *IEEE Trans. Energy Convers.*, vol. 31, no. 3, pp. 1002–1011, 2016, doi: 10.1109/TEC.2016.2558631.

[37] B. Mohammadpour, M. Zareie, S. Eren, and M. Pahlevani, "Stability analysis of the slip mode frequency shift islanding detection in single phase PV inverters," *IEEE Int. Symp. Ind. Electron.*, pp. 873–878, 2017, doi: 10.1109/ISIE.2017.8001361.

[38] S. Akhlaghi, A. Akhlaghi, and A. A. Ghadimi, "Performance analysis of the Slip mode frequency shift islanding detection method under different inverter interface control strategies," *2016 IEEE Power Energy Conf. Illinois, PECI 2016*, pp. 1–7, 2016, doi: 10.1109/PECI.2016.7459250.

[39] M. E. Ropp, M. Begovic, A. Rohatgi, G. A. Kern, R. H. Bonn, and S. Gonzalez, "Determining the relative effectiveness of islanding detection methods using phase criteria and nondetection zones," *IEEE Trans. Energy Convers.*, vol. 15, no. 3, pp. 290–296, 2000, doi: 10.1109/60.875495.

[40] W. bower, *"Evaluation of Islanding Detection Methods for Photovoltaic Utility-Interactive Power Systems,"* Report IEA PVPS T5-09, 2002.

[41] E. Vazquez, N. Vazquez, and R. Femat, "Modified Sandia voltage shift anti-islanding scheme for distributed power generator systems," *IET Power Electron.*, vol. 13, no. 18, pp. 4226–4234, 2020, doi: 10.1049/iet-pel.2020.0735.

[42] U. Bartoccini, G. Barchi, and E. Nunzi, "Methods and tools for frequency jump detection," *2009 IEEE Int. Work. Adv. Methods Uncertain. Estim. Meas. AMUEM 2009*, no. July, pp. 109–112, 2009, doi: 10.1109/AMUEM.2009.5207593.

[43] L. Galleani and P. Tavella, "Robust detection of fast and slow frequency jumps of atomic clocks," *IEEE Trans. Ultrason. Ferroelectr. Freq. Control*, vol. 64, no. 2, pp. 475–485, 2017, doi: 10.1109/TUFFC.2016.2625311.

[44] W. J. Chiang, H. L. Jou, and J. C. Wu, "Active islanding detection method for inverter-based distribution generation power system," *Int. J. Electr. Power Energy Syst.*, vol. 42, no. 1, pp. 158–166, 2012, doi: 10.1016/j.ijepes.2012.03.038.

[45] G. Hernandez-Gonzalez and R. Iravani, "Current injection for active islanding detection of electronically-interfaced distributed resources," *IEEE Trans. Power Deliv.*, vol. 21, no. 3, pp. 1698–1705, Jul. 2006, doi: 10.1109/TPWRD.2006.876980.

[46] K. N. E. Ku Ahmad, J. Selvaraj, and N. A. Rahim, "A review of the islanding detection methods in grid-connected PV inverters," *Renew. Sustain. Energy Rev.*, vol. 21, pp. 756–766, May 2013, doi: 10.1016/j.rser.2013.01.018.

[47] G. Chicco, J. Schlabbach, and F. Spertino, "Operation of multiple inverters in grid-connected large-size photovoltaic installations," in *Proceedings International Conference and Exhibition on Electricity Distribution—Part 1, (CIRED)*, IET, 2009, pp. 1–4.

[48] J.-H. J.-G. Kim, J.-H. J.-G. Kim, Y.-H. Ji, Y.-C. Jung, and C.-Y. Won, "An islanding detection method for a grid-connected system based on the Goertzel algorithm," *IEEE Trans. Power Electron.*, vol. 26, no. 4, pp. 1049–1055, Apr. 2011, doi: 10.1109/TPEL.2011.2107751.

[49] X. Wang, W. Freitas, W. Xu, and V. Dinavahi, "Impact of DG interface controls on the Sandia frequency shift antiislanding method," *IEEE Trans. Energy Convers.*, vol. 22, no. 3, pp. 792–794, Sep. 2007, doi: 10.1109/TEC.2007.902668.

[50] Wilsun Xu. Method for identifying a system anomaly in a power distribution system. 10598736. United States: US Patent Application, issued 2007.

[51] S. Perlenfein, M. Ropp, J. Neely, S. Gonzalez, and L. Rashkin, "Subharmonic power line carrier (PLC) based island detection," in *Conference Proceedings – IEEE Applied Power Electronics Conference and Exposition – APEC*, vol. 2015, no. May, pp. 2230–2236, 2015, doi: 10.1109/APEC.2015.7104659.

[52] W. Xu et al., "A power line signaling based technique for anti-islanding protection of distributed generators—part I: Scheme and analysis," *IEEE Trans. Power Deliv.*, vol. 22, no. 3, pp. 1758–1766, Jul. 2007, doi: 10.1109/TPWRD.2007.899618.

[53] Y. Zhang, Y. Xu, and Z. Y. Dong, "Robust ensemble data analytics for incomplete PMU measurements-based power system stability assessment," *IEEE Trans. Power Syst.*, vol. 33, no. 1, pp. 1124–1126, 2017, doi: 10.1109/tpwrs.2017.2698239.

[54] S. Wang, P. Dehghanian, and L. Li, "Power grid online surveillance through PMU-embedded convolutional neural networks," *IEEE Trans. Ind. Appl.*, vol. 56, no. 2, pp. 1146–1155, 2020, doi: 10.1109/TIA.2019.2958786.

[55] G. Bayrak, "A remote islanding detection and control strategy for photovoltaic-based distributed generation systems," *Energy Convers. Manag.*, vol. 96, pp. 228–241, 2015, doi: 10.1016/j.enconman.2015.03.004.

[56] C. Naradon, C. I. Chai, M. Leelajindakrairerk, and C. I. Chow, "A case study on the interoperability of the Direct Transfer Trip (DTT) technique with carrier signal protection schemes (PTT and DEF) and SCADA system between two utilities in Thailand," in *Conference Proceedings – 2017 17th IEEE International Conference on Environment and Electrical Engineering and 2017 1st IEEE Industrial and Commercial Power Systems Europe (EEEIC / I&CPS Europe)*, IEEE, 2017, doi: 10.1109/EEEIC.2017.7977555.

[57] G. Artale et al., "A new PLC-based smart metering architecture for medium/low voltage grids: Feasibility and experimental characterization," *Measurement*, vol. 129, pp. 479–488, Dec. 2018, doi: 10.1016/j.measurement.2018.07.070.

[58] C. M. Riley, B. K. Lin, T. G. Habetler, and G. B. Kliman, "Stator current harmonics and their causal vibrations: A preliminary investigation of sensorless vibration monitoring applications," *IEEE Trans. Ind. Appl.*, vol. 35, no. 1, pp. 94–99, 1999, doi: 10.1109/28.740850.

[59] T. S. Ustun, S. M. Farooq, and S. M. S. Hussain, "Implementing secure routable GOOSE and SV messages based on IEC 61850-90-5," *IEEE Access*, vol. 8, pp. 26162–26171, 2020, doi: 10.1109/ACCESS.2020.2971011.

[60] P. A. Pegoraro, K. Brady, P. Castello, C. Muscas, and A. von Meier, "Compensation of systematic measurement errors in a PMU-based monitoring system for electric distribution grids," *IEEE Trans. Instrum. Meas.*, vol. 68, no. 10, pp. 3871–3882, Oct. 2019, doi: 10.1109/TIM.2019.2908703.

[61] S. Sarangi and A. K. Pradhan, "Adaptive direct underreaching transfer trip protection scheme for the three-terminal line," *IEEE Trans. Power Deliv.*, vol. 30, no. 6, pp. 2383–2391, Dec. 2015, doi: 10.1109/TPWRD.2015.2388798.

[62] S. Dutta, P. K. Sadhu, M. Jaya Bharata Reddy, and D. K. Mohanta, "Shifting of research trends in islanding detection method – a comprehensive survey," *Prot. Control Mod. Power Syst.*, vol. 3, no. 1, p. 1, Dec. 2018, doi: 10.1186/s41601-017-0075-8.

[63] K. Shi, H. Ye, P. Xu, Y. Yang, and F. Blaabjerg, "An islanding detection based on droop characteristic for virtual synchronous generator," *Int. J. Electr. Power Energy Syst.*, vol. 123, no. August 2019, p. 106277, 2020, doi: 10.1016/j.ijepes.2020.106277.

[64] E. Serban, C. Pondiche, and M. Ordonez, "Islanding detection search sequence for distributed power generators under AC grid faults," *IEEE Trans. Power Electron.*, vol. 30, no. 6, pp. 3106–3121, 2015, doi: 10.1109/TPEL.2014.2331685.

[65] J. Zhang, D. Xu, G. Shen, Y. Zhu, N. He, and J. Ma, "An improved islanding detection method for a grid-connected inverter with intermittent bilateral reactive power variation," *IEEE Trans. Power Electron.*, vol. 28, no. 1, pp. 268–278, 2013, doi: 10.1109/TPEL.2012.2196713.

[66] R. Sirjani and C. F. Okwose, "Combining two techniques to develop a novel islanding detection method for distributed generation units," *Meas. J. Int. Meas. Confed.*, vol. 81, pp. 66–79, 2016, doi: 10.1016/j.measurement.2015.12.001.

[67] X. Chen and Y. Li, "An islanding detection algorithm for inverter-based distributed generation based on reactive power control," *IEEE Trans. Power Electron.*, vol. 29, no. 9, pp. 4672–4683, 2014, doi: 10.1109/TPEL.2013.2284236.

[68] A. M. Nayak, M. Mishra, and B. B. Pati, "A hybrid islanding detection method considering voltage unbalance factor," in *2020 IEEE International Symposium on Sustainable Energy, Signal Processing and Cyber Security (ISSSC)*, Dec. 2020, pp. 1–5, doi: 10.1109/iSSSC50941.2020.9358881.

[69] B. Sun, J. Mei, and J. Zheng, "A novel islanding detection method based on positive feedback between active current and voltage unbalance factor," in *2014 IEEE Innovative Smart Grid Technologies – Asia, ISGT ASIA 2014*, no. 1, pp. 31–34, 2014, doi: 10.1109/ISGT-Asia.2014.6873759.

[70] M. Khodaparastan, H. Vahedi, F. Khazaeli, and H. Oraee, "A novel hybrid islanding detection method for inverter-based DGs using SFS and ROCOF," *IEEE Trans. Power Deliv.*, vol. 32, no. 5, pp. 2162–2170, Oct. 2017, doi: 10.1109/TPWRD.2015.2406577.

[71] P. Mahat, Z. Chen, and B. Bak-Jensen, "Review of islanding detection methods for distributed generation," in *2008 Third International Conference on Electric Utility Deregulation and Restructuring and Power Technologies*, Apr. 2008, pp. 2743–2748, doi: 10.1109/DRPT.2008.4523877.

[72] D. Kumar, "Islanding detection in microgrid compromising missing values Using NI Sensors," *IEEE Syst. J.*, pp. 1–10, 2021, doi: 10.1109/JSYST.2021.3055566.

[73] H. Samet, F. Hashemi, and T. Ghanbari, "Minimum non detection zone for islanding detection using an optimal Artificial Neural Network algorithm based on PSO," *Renew. Sustain. Energy Rev.*, vol. 52, pp. 1–18, Dec. 2015, doi: 10.1016/j.rser.2015.07.080.

[74] F. Hashemi, N. Ghadimi, and B. Sobhani, "Islanding detection for inverter-based DG coupled with using an adaptive neuro-fuzzy inference system," *Int. J. Electr. Power Energy Syst.*, vol. 45, no. 1, pp. 443–455, 2013, doi: 10.1016/j.ijepes.2012.09.008.

[75] J. R. Zhang, J. Zhang, T. M. Lok, and M. R. Lyu, "A hybrid particle swarm optimization-back-propagation algorithm for feedforward neural network training," *Appl. Math. Comput.*, vol. 185, no. 2, pp. 1026–1037, 2007, doi: 10.1016/j.amc.2006.07.025.

[76] G. Das, P. K. Pattnaik, and S. K. Padhy, "Artificial neural network trained by particle Swarm optimization for non-linear channel equalization," *Expert Syst. Appl.*, vol. 41, no. 7, pp. 3491–3496, 2014, doi: 10.1016/j.eswa.2013.10.053.

[77] K. H. Chao, C. L. Chiu, C. J. Li, and Y. C. Chang, "A novel neural network with simple learning algorithm for islanding phenomenon detection of photovoltaic systems," *Expert Syst. Appl.*, vol. 38, no. 10, pp. 12107–12115, 2011, doi: 10.1016/j.eswa.2011.02.175.

[78] A. Moeini, A. Darabi, S. M. R. Rafiei, and M. Karimi, "Intelligent islanding detection of a synchronous distributed generation using governor signal clustering," *Electr. Power Syst. Res.*, vol. 81, no. 2, pp. 608–616, 2011, doi: 10.1016/j.epsr.2010.10.023.

[79] D. Mlakic, H. R. Baghaee, and S. Nikolovski, "A novel ANFIS-based islanding detection for inverter-interfaced microgrids," *IEEE Trans. Smart Grid*, vol. 10, no. 4, pp. 4411–4424, Jul. 2019, doi: 10.1109/TSG.2018.2859360.

[80] M. A. Khan, V. S. Bharath Kurukuru, A. Haque, and S. Mekhilef, "Islanding classification mechanism for grid-connected photovoltaic systems," *IEEE J. Emerg. Sel. Top. Power Electron.*, vol. 9, no. 2, pp. 1966–1975, Apr. 2021, doi: 10.1109/JESTPE.2020.2986262.

[81] M. A. Khan, A. Haque, and V. S. B. Kurukuru, "Machine learning based islanding detection for grid connected photovoltaic system," in *2019 International Conference on Power Electronics, Control and Automation (ICPECA)*, Nov. 2019, pp. 1–6, doi: 10.1109/ICPECA47973.2019.8975614.

[82] A. Fatama, A. Haque, and M. A. Khan, "A multi feature based islanding classification technique for distributed generation systems," in *2019 International Conference on Machine Learning, Big Data, Cloud and Parallel Computing (COMITCon)*, Feb. 2019, pp. 160–166, doi: 10.1109/COMITCon.2019.8862442.

[83] M. A. Khan, A. Haque, and V. S. B. Kurukuru, "An efficient islanding classification technique for single phase grid connected photovoltaic system," in *2019 International Conference on Computer and Information Sciences (ICCIS)*, Apr. 2019, pp. 1–6, doi: 10.1109/ICCISci.2019.8716438.

[84] O. N. Faqhruldin, S. Member, E. F. El-saadany, S. Member, H. H. Zeineldin, and S. Member, "A universal islanding detection technique for distributed generation using pattern recognition," *IEEE Trans. Smart Grid*, vol. 5, no. 4, pp. 1985–1992, 2014.

[85] B. Matic-Cuka and M. Kezunovic, "Islanding detection for inverter-based distributed generation using support vector machine method," *IEEE Trans. Smart Grid*, vol. 5, no. 6, pp. 2676–2686, Nov. 2014, doi: 10.1109/TSG.2014.2338736.

[86] S. D. Kermany, M. Joorabian, S. Deilami, and M. A. S. Masoum, "Hybrid islanding detection in microgrid with multiple connection points to smart grids using fuzzy-neural network," *IEEE Trans. Power Syst.*, vol. 32, no. 4, pp. 2640–2651, 2017, doi: 10.1109/TPWRS.2016.2617344.

[87] C. Hsieh, J. Lin, and S. Huang, "Electrical power and energy systems enhancement of islanding-detection of distributed generation systems via wavelet transform-based approaches," *Int. J. Electr. Power Energy Syst.*, vol. 30, no. 10, pp. 575–580, 2008, doi: 10.1016/j.ijepes.2008.08.006.

[88] H. Shayeghi, B. Sobhani, E. Shahryari, and A. Akbarimajd, "Optimal neuro-fuzzy based islanding detection method for distributed generation," *Neurocomputing*, vol. 177, pp. 478–488, 2016, doi: 10.1016/j.neucom.2015.11.056.

[89] H. T. Do, X. Zhang, N. V. Nguyen, S. S. Li, and T. T. T. Chu, "Passive-islanding detection method using the wavelet packet transform in grid-connected photovoltaic systems," *IEEE Trans. Power Electron.*, vol. 31, no. 10, pp. 6955–6967, 2016, doi: 10.1109/TPEL.2015.2506464.

[90] S. C. Paiva, R. L. de A. Ribeiro, D. K. Alves, F. B. Costa, and T. de O. A. Rocha, "A wavelet-based hybrid islanding detection system applied for distributed generators interconnected to AC microgrids," *Int. J. Electr. Power Energy Syst.*, vol. 121, no. March, p. 106032, 2020, doi: 10.1016/j.ijepes.2020.106032.

[91] M. Ahmadipour, H. Hizam, M. Lutfi Othman, M. A. M. Radzi, and N. Chireh, "A novel islanding detection technique using modified Slantlet transform in multi-distributed generation," *Int. J. Electr. Power Energy Syst.*, vol. 112, no. April, pp. 460–475, 2019, doi: 10.1016/j.ijepes.2019.05.008.

[92] P. K. K. Dash, M. Padhee, and T. K. K. Panigrahi, "A hybrid time–frequency approach based fuzzy logic system for power island detection in grid connected distributed generation," *Int. J. Electr. Power Energy Syst.*, vol. 42, no. 1, pp. 453–464, Nov. 2012, doi: 10.1016/j.ijepes.2012.04.003.

[93] T. S. Menezes, R. A. S. Fernandes, and D. V. Coury, "Intelligent islanding detection with grid topology adaptation and minimum non-detection zone," *Electr. Power Syst. Res.*, vol. 187, no. February, p. 106470, 2020, doi: 10.1016/j.epsr.2020.106470.

[94] M. Ahmadipour, H. Hizam, M. L. Othman, M. A. M. Radzi, and A. S. Murthy, "Islanding detection technique using Slantlet transform and ridgelet probabilistic neural network in grid-connected photovoltaic system," *Appl. Energy*, vol. 231, no. August, pp. 645–659, Dec. 2018, doi: 10.1016/j.apenergy.2018.09.145.

[95] M. A. Farhan and K. Shanti Swarup, "Mathematical morphology-based islanding detection for distributed generation," *IET Gener. Transm. Distrib.*, vol. 10, no. 2, pp. 518–525, 2016, doi: 10.1049/iet-gtd.2015.0910.

[96] S. Gautam and S. M. Brahma, "Overview of mathematical morphology in power systems – A tutorial approach," *2009 IEEE Power Energy Society General Meeting PES '09*, IEEE, pp. 1–7, 2009, doi: 10.1109/PES.2009.5275190.

[97] A. Gururani, S. R. Mohanty, and J. C. Mohanta, "Microgrid protection using Hilbert-Huang transform based-differential scheme," *IET Gener. Transm. Distrib.*, vol. 10, no. 15, pp. 3707–3716, 2016, doi: 10.1049/iet-gtd.2015.1563.

[98] B. K. Chaitanya and A. Yadav, "Hilbert-huang transform based islanding detection scheme for Distributed generation," *8th IEEE Power India International Conference PIICON 2018*, pp. 3–7, 2018, doi: 10.1109/POWERI.2018.8704444.

[99] M. Mishra, M. Sahani, and P. K. Rout, "An islanding detection algorithm for distributed generation based on Hilbert–Huang transform and extreme learning machine," *Sustain. Energy Grids Networks*, vol. 9, pp. 13–26, 2017, doi: 10.1016/j.segan.2016.11.002.

[100] Y. Guo, K. Li, D. M. Laverty, and Y. Xue, "Synchrophasor-based islanding detection for distributed generation systems using systematic principal component analysis approaches," *IEEE Trans. Power Deliv.*, vol. 30, no. 6, pp. 2544–2552, 2015, doi: 10.1109/TPWRD.2015.2435158.

[101] X. Liu, D. M. Laverty, R. J. Best, K. Li, D. J. Morrow, and S. McLoone, "Principal component analysis of wide-area phasor measurements for islanding detection – A geometric view," *IEEE Trans. Power Deliv.*, vol. 30, no. 2, pp. 976–985, 2015, doi: 10.1109/TPWRD.2014.2348557.

[102] M. Padhee, P. K. Dash, K. R. Krishnanand, and P. K. Rout, "A fast gauss-newton algorithm for islanding detection in distributed generation," *IEEE Trans. Smart Grid*, vol. 3, no. 3, pp. 1181–1191, 2012, doi: 10.1109/TSG.2012.2199140.

[103] S. Y. Xue and S. X. Yang, "Power system frequency estimation using supervised Gauss-Newton algorithm," *Meas. J. Int. Meas. Confed.*, vol. 42, no. 1, pp. 28–37, 2009, doi: 10.1016/j.measurement.2008.03.018.

[104] A. Taheri Kolli and N. Ghaffarzadeh, "A novel phaselet-based approach for islanding detection in inverter-based distributed generation systems," *Electr. Power Syst. Res.*, vol. 182, no. September 2019, 2020, doi: 10.1016/j.epsr.2020.106226.

[105] E. Rosolowski, A. Burek, and L. Jedut, *A New Method for Islanding Detection in Distributed Generation*, Wroclaw University of Technology, Wroclaw, Poland, 2007.

[106] S. R. Samantaray, K. El-Arroudi, G. Joos, and I. Kamwa, "A fuzzy rule-based approach for islanding detection in distributed generation," *IEEE Trans. Power Deliv.*, vol. 25, no. 3, pp. 1427–1433, Jul. 2010, doi: 10.1109/TPWRD.2010.2042625.

[107] M. A. Khan, A. Haque, and V. S. B. Kurukuru, "Performance assessment of stand-alone transformerless inverters," *Int. Trans. Electr. Energy Syst.*, vol. 30, no. 1, pp. 1–20, Jan. 2020, doi: 10.1002/2050-7038.12156.

[108] M. Elnozahy, E. El-saadany, and M. Salama, "A Robust Wavelet-ANN based technique for islanding detection," in *Power and Energy Society General Meeting*, 2011, pp. 1–8.

[109] V. S. B. Kurukuru, F. Blaabjerg, M. A. Khan, and A. Haque, "A novel fault classification approach for photovoltaic systems," *Energies*, vol. 13, no. 2, p. 308, Jan. 2020, doi: 10.3390/en13020308.

[110] Y. Li, N. Lu, X. Wang, and B. Jiang, "Islanding fault detection based on data-driven approach with active developed reactive power variation," *Neurocomputing*, vol. 337, pp. 97–109, 2019, doi: 10.1016/j.neucom.2019.01.054.

[111] X. Kong, X. Xu, Z. Yan, S. Chen, H. Yang, and D. Han, "Deep learning hybrid method for islanding detection in distributed generation," *Appl. Energy*, vol. 210, no. July, pp. 776–785, 2018, doi: 10.1016/j.apenergy.2017.08.014.

[112] Goverment of India, "*National Electrical Code*," 2011.

[113] International Electrotechnical Commission, "*IEC 62116-2014*," 2014.

[114] North American Electric Reliability Corporation (NERC), "*1200 MW Fault Induced Solar Photovoltaic Resource Interruption Disturbance Report*," no. June, p. 32, 2017.

[115] A. Johnson, "Fault Ride Through RfG - Compliance," 2016. www2.nationalgrid.com/WorkArea/DownloadAsset.aspx?id=33095 (accessed Dec. 21, 2022).

[116] F. Blaabjerg, F. Iov, R. Teodorescu, and Z. Chen, "Power electronics in renewable energy systems," in *2006 12th International Power Electronics and Motion Control Conference*, Aug. 2006, pp. 1–17, doi: 10.1109/EPEPEMC.2006.4778368.

[117] F. G. Baum, "Voltage regulation and insulation for large power long distance transmission systems," *Trans. Am. Inst. Electr. Eng.*, vol. XL, pp. 1017–1077, Jan. 1921, doi: 10.1109/T-AIEE.1921.5060737.

[118] R. Teodorescu, M. Liserre, and P. Rodríguez, *Grid Converters for Photovoltaic and Wind Power Systems*. John Wiley & Sons, Ltd., 2011.

[119] W. A. Omran, M. Kazerani, and M. M. A. Salama, "Investigation of methods for reduction of power fluctuations generated from large grid-connected photovoltaic systems," *IEEE Trans. Energy Convers.*, vol. 26, no. 1, pp. 318–327, Mar. 2011, doi: 10.1109/TEC.2010.2062515.

[120] J. M. Guerrero, "Microgrids: Integration of distributed energy resources into the smart-grid," in *2010 IEEE International Symposium on Industrial Electronics*, Jul. 2010, pp. 4281–4414, doi: 10.1109/ISIE.2010.5637667.

[121] J. Guerrero et al., "Distributed generation: Toward a new energy paradigm," *IEEE Ind. Electron. Mag.*, vol. 4, no. 1, pp. 52–64, Mar. 2010, doi: 10.1109/MIE.2010.935862.

[122] A. Núñez-Reyes, D. Marcos Rodríguez, C. Bordons Alba, and M. Á. Ridao Carlini, "Optimal scheduling of grid-connected PV plants with energy storage for integration in the electricity market," *Sol. Energy*, vol. 144, pp. 502–516, Mar. 2017, doi: 10.1016/j.solener.2016.12.034.

[123] D. Arcos-Aviles, J. Pascual, L. Marroyo, P. Sanchis, and F. Guinjoan, "Fuzzy logic-based energy management system design for residential grid-connected microgrids," *IEEE Trans. Smart Grid*, vol. 9, no. 2, pp. 530–543, Mar. 2018, doi: 10.1109/TSG.2016.2555245.

[124] D. E. Olivares, C. A. Canizares, and M. Kazerani, "A centralized energy management system for isolated microgrids," *IEEE Trans. Smart Grid*, vol. 5, no. 4, pp. 1864–1875, Jul. 2014, doi: 10.1109/TSG.2013.2294187.

[125] Z. Li, J. Huang, B. Y. Liaw, and J. Zhang, "On state-of-charge determination for lithium-ion batteries," *J. Power Sources*, vol. 348, pp. 281–301, Apr. 2017, doi: 10.1016/j.jpowsour.2017.03.001.

[126] R. H. Byrne, T. A. Nguyen, D. A. Copp, B. R. Chalamala, and I. Gyuk, "Energy management and optimization methods for grid energy storage systems," *IEEE Access*, vol. 6, pp. 13231–13260, 2018, doi: 10.1109/ACCESS.2017.2741578.

[127] M. A. Hannan et al., "A review of internet of energy based building energy management systems: Issues and recommendations," *IEEE Access*, vol. 6, pp. 38997–39014, 2018, doi: 10.1109/ACCESS.2018.2852811.

[128] A. Aktas, K. Erhan, S. Ozdemir, and E. Ozdemir, "Experimental investigation of a new smart energy management algorithm for a hybrid energy storage system in smart grid applications," *Electr. Power Syst. Res.*, vol. 144, pp. 185–196, Mar. 2017, doi: 10.1016/j.epsr.2016.11.022.

[129] M. N. Arafat, S. Palle, Y. Sozer, and I. Husain, "Transition control strategy between Standalone and grid-connected operations of voltage-source inverters," *IEEE Trans. Ind. Appl.*, vol. 48, no. 5, pp. 1516–1525, Sep. 2012, doi: 10.1109/TIA.2012.2210013.

[130] J. Machowski, W. J. Bialek, and R. J. Bumby, *Power System Dynamics: Stability and Control*. John Wiley & Sons, Ltd, 2008.

Index

Note: **Bold** page numbers refer to tables and *italic* page numbers refer to figures.

For Product Safety Concerns and Information please contact our EU
representative GPSR@taylorandfrancis.com
Taylor & Francis Verlag GmbH, Kaufingerstraße 24, 80331 München, Germany

www.ingramcontent.com/pod-product-compliance
Lightning Source LLC
Chambersburg PA
CBHW060403220326
41598CB00023B/3008